T0189962

Studies in Computational Intelligence

Data, Semantics and Cloud Computing

Volume 485

Series Editor

J. Kacprzyk, Warsaw, Poland

For further volumes:
http://www.springer.com/series/11756

Sarinder K. Dhillon · Amandeep S. Sidhu

Data Intensive Computing for Biodiversity

Dr. Sarinder K. Dhillon
Faculty of Science
Institute of Biological Sciences
University of Malaya
Kuala Lumpur
Malaysia

Dr. Amandeep S. Sidhu
Curtin Sarawak Research Institute
Curtin University
Miri, Sarawak
Malaysia

ISSN 1860-949X ISSN 1860-9503 (electronic)
ISBN 978-3-642-44189-9 ISBN 978-3-642-38047-1 (eBook)
DOI 10.1007/978-3-642-38047-1
Springer Heidelberg New York Dordrecht London

Softcover re-print of the Hardcover 1st edition 2013

Printed on acid-free paper

Springer is part of Springer Science+Business Media (www.springer.com)

Preface

Biodiversity information has grown tremendously over these years, so has the need to digitize, store, and share the diverse knowledge in this area. Numerous data integration efforts have been researched on and built for various biological data sets; however, in Malaysia there is considerable need in sharing and integration of biodiversity data. This is due to issues such as (i) some institutions have information documented in the form of databases, while other institutions are looking into it; (ii) data stored in heterogeneous formats; (iii) many databases are still offline; (iv) many databases are still unstructured; (v) databases are still independent of each other; and (vi) no uniformity in vocabulary used to manage data. Therefore, this book focused on the development of a data integration framework for retrieval of biodiversity information from heterogeneous and distributed data sources.

The data integration system proposed in this book links remote databases in a networked environment, supports heterogeneous databases and data formats, links databases hosted on multiple platforms, and provides data security for database owners by allowing them to keep and maintain their own data and to choose information to be shared and linked.

Chapter 1 provides an overview of biodiversity information, the aim and justifications for this research. Chapter 2 describes the preliminary study whereby an existing integration system, DiGIR, was used to integrated biodiversity databases. This chapter also presents the results of DiGIR implementation and discusses the need for a new database integration system. Chapter 3 contains the literature review which consists of research on existing database integration systems, analysis of related components, technologies, and technical aspects of database integration. The findings in this chapter were used to build the database integration system in Chapter 4. In Chapter 4, the methods and materials concerning the domain of research are presented. The following chapters, (Chapters 5 and 6), contain the results obtained from this study. Chapter 5 describes the biodiversity data format used to build the relational biodiversity databases. Chapter 6 describes the results of the proposed database integration system. The discussion in Chapter 7 reflects the final framework of the biodiversity database integration system.

Kuala Lumpur, Malaysia, March 2013

Sarinder K. Dhillon
Amandeep S. Sidhu

Contents

Chapter 1
Introduction

1.1 Overview

Bioinformatics deals with computational management and analysis of biological information (Buttler et al. 2002; Nilges et al. 2002).

Two key words in the above definition are biological and information which form the basis of this book. 'Sir Robert May, the Chief Scientist for the United Kingdom and now President of the Royal Society of London, stated in 1998, "*…there will be winners and there will be losers… The next century will be the 'Age of Biology', just as [the twentieth century was] the age of physics and astronomy. Specifically, those countries who best know how to correlate, analyze and communicate biological information will be in the leading position to achieve economic and scientific advances.*"

Biological information is a huge subject in the biological research as the term biological consists of many areas such as medical, genetics, biodiversity, environment, biotechnology and many more. However, this book is focused on biodiversity information.

1.2 Global Perspective on Biodiversity Information

The Convention on Biological Diversity (CBD 2005) stated that "*biological diversity—or biodiversity—is the term given to the variety of life on Earth and the natural patterns it forms. The biodiversity seen today is the fruit of billions of years of evolution, shaped by natural processes and, increasingly, by the influence of humans. It forms the Web of life of which humans are an integral part and upon which they so fully depend. It is the combination of Life forms and their interactions with each other and with the rest of the environment that has made Earth a uniquely habitable place for humans. Biodiversity provides a large number of goods and services that sustain our lives*". At the 1992 Earth Summit in Rio de Janeiro, it was agreed that a comprehensive strategy is needed for sustainability as

S. K. Dhillon and A. S. Sidhu, *Data Intensive Computing for Biodiversity*,
Data, Semantics and Cloud Computing 485, DOI: 10.1007/978-3-642-38047-1_1,

it is highly important for future generations to have a healthy environment. In line with this, many organisations in the world have set out commitments to achieve this along with economic development (CBD, 2005).

During the last decade there has been an increasing interest in gathering and analyzing biodiversity information for the scientific administration of natural resources. Many initiatives around the globe arose to sustain the biodiversity resources and communicate biological information using computational tools. Examples include the Canadian Clearing-House Mechanism (Secretariat on the Convention on Biological Diversity 2005), the Japan's Global Taxonomy Initiative (GTI) (Ando and Watanabe 2003), Inter-American Biodiversity Information Network (IABIN 2004a) and Global Biodiversity Information Facility (GBIF 2004).

Therefore, as predicted by Sir Robert May, the world is moving towards the direction of sustaining and communicating biological information as the importance is becoming more and more transparent.

However, as far as communication is concerned, biodiversity information clusters are usually disperse, unreported and in some cases inaccessible. There is an emerging need to overcome this setback by looking at issues concerning communication of biodiversity information (biodiversity informatics). There is a global need to link all the disperse information clusters in remote and distributed medium.

Technological advances within distributed network communications for biodiversity informatics offer a window of opportunity for gathering and maintaining information repositories about biodiversity, analyzing this information, reporting and visualizing it (Blum 2000). However, poor results have been obtained so far (Schnase et al. 2003). This is due to: (a) each researcher manages data according to their own preference; (b) collaboration between the biologists and the computer scientists is still very little; (c) the lack of a common or standard format adopted for data representation, makes it difficult to integrate data; (d) some institutions or even individuals are reluctant to publish or share the collected data.

Reflecting the concerns discussed above, it can be concluded that there is a need to integrate the different views and versions of taxonomic data, making it available in simple formats, with friendly interfaces to be shared among the scientific community and to bring them together to work as a team to achieve their respective goals, without expecting them to export their collected information to a centralized data warehouse.

1.3 Malaysian Perspective of Biodiversity Information

Malaysia has been identified as one of the world's twelve mega-diversity areas with extreme rich biological resources. There are over 15,000 known species of higher plants, 300 species of mammals, 254 species of breeding birds, 198 species of amphibians, 379 species of reptiles, over 150,000 species of the invertebrates, and over 4,000 species of marine fishes and 449 species of freshwater fishes in Malaysia (Burhanuddin 2000).

Despite having such rich biodiversity resources, Malaysia does not have a central physical body for storing natural history collections (see Sect. 2.3.3 for details) while the virtual repositories are maintained by individuals or organizations disparately. This is explained further in the following sections.

1.3.1 Physical Repositories

University of Malaya (UM) was established on 8th October 1949 as a national institution to serve the higher education needs of the Federation of Malaya and Singapore (Formerly known as Raffles University). The growth of the University was very rapid during the first decade of its establishment and resulted in the setting up of two autonomous divisions in 1959, one located in Singapore (later becoming the National University of Singapore) and the other in Kuala Lumpur (retaining the name University of Malaya). Today, National University of Singapore (NUS) still maintains the Raffles Museum (Sivasothi 2006) which was constructed back in 1849. Due to the historical background, it also contains specimens collected in Malaysia. UM has "Rimba Ilmu" Botanical Garden (Wong 2005) and Zoology Museum which are two important physical bodies that store the vast biodiversity specimens collected by scientists in the university.

Besides Raffles Museum, some of the Malaysian (especially Sarawak) specimens are kept in Natural History Museum in London and Natural History Museum at Tring. Sarawak is famous with its rich diversity of biological specimens which date back to the time of A.R Wallace. Wallace, who arrived in Sarawak at the invitation of Rajah Sir James Brooke on Nov 1 1854, spent fifteen months exploring and collecting an enormous 25,000 specimens, including 2,000 beetle species, 1,500 moth species and 1,500 other insect orders along the Sarawak River valley from Santubong to Bau as well as the peat swamps of Simunjan. The collections, which he sold to private collectors and institutions in the United Kingdom to finance his travels in the region, are now kept at the Natural History Museum in London and Natural History Museum at Tring.

Besides UM, University Science Malaysia (USM) has also taken steps towards research in biodiversity. It adopted the concept of "the university in a garden" to promote the preservation of green areas as integral to the development of the intellect and thus enhancing the spirit and practice of nature conservation.

University Putra Malaysia (UPM) too plays a very active role in biodiversity conservation in the country. The Institute of BioScience in UPM is a center of excellence for biological research, including biodiversity.

The Institute of Medical Research (IMR 2006), a deserving institute which existed for about 112 years now, has been actively involved in biodiversity research especially medical related organisms, bacteria, virus, protozoas, parasites and pathogens. It is one of the oldest institutions in Malaysia which has a physical repository of specimens.

Other institutions, for example FRIM (Forest Research Institute of Malaysia) and MARDI (Malaysian Agricultural Research and Development Institute) have some amount of biodiversity collections in Malaysia.

The above organizations and institutions explain about the physical distributed data warehouses that records, stores and analyzes biodiversity information. However, these physical entities work individually with minimum collaboration among each other.

1.3.2 Virtual Repositories

While the physical repositories are essential to store the various biodiversity specimens, the virtual repositories are equally important. They can be used to manage, analyze share and disseminate the data in a more structured way.

In Malaysia, several initiatives have been undertaken to digitize their biodiversity information. Some of these initiatives are described below.

In 2003, the Institute of Biological Sciences at UM started an initiative named Integrated Biological Sciences Initiative (IBDI) (Sarinder et al. 2005), in which relational biodiversity databases were developed and museum collections catalogued. This initiative resulted in the digitization and subsequent electronic availability of vast amount of biodiversity data in UM.

Biodiversity Databases for Malaysian Flora and Fauna (Napis et al. 2001) has been an important effort to catalogue some of the rich biodiversity resources in Malaysia. Palm Oil Research Institute of Malaysia (PORIM) maintains an oil palm database, accessible to registered internet users only. In addition, the Forest Research Institute of Malaysia (FRIM) provides Web users limited database access to its huge forest resource collections (Merican et al. 2002).

1.3.3 Problems Concerning Biodiversity Information in Malaysia

From the above, it can be said that a wealth of information exists on Malaysia's biodiversity resources and associated knowledge. This may be in form of specimens, gray literature such as unpublished reports, and books, monographs and scientific papers and electronic databases.

Undoubtedly, the problem faced by Malaysia in biodiversity information is basically one of dissemination and hence retrieval of information in a networked environment. Presently there is no system to link the scattered data so as to facilitate exchange of data amongst the different databases available in the country.

The data security issues faced by researchers have been seen as issues which require immediate attention.

Besides the above, there are other impediments which can be summarized as follows:

1. To date, only some institutions have information documented in the form of databases.
2. Databases are in heterogeneous formats.
3. Very few databases are online.
4. Only some of these databases are well-structured.
5. These databases exist independently.
6. There is no meta-data (data- dictionary).

As a result, there is a challenge to tackle the concerns on biodiversity information dissemination and retrieval, which, if they persist, will become a hindrance for Malaysia to be in the leading position to achieve economic and scientific advances, as clearly quoted by Sir Robert May.

References

Ando, K., & Watanabe, M. (2003). *Global taxonomy initiative (GTI) and taxonomy*. Gaithersburg: National Institute of Evaluation and Technology and National Institute of Environment. Tsukuba, Ibaraki Prefecture, Japan.

Blum, S. (2000). Overview of biodiversity informatics (online). Available from: http://www.calacademy.org/research/informatics/sblum/pub/biodiv_informatics.html Accessed 12 Feb 2006.

Burhanuddin, M. (2000). Biodiversity and information in Malaysia (online). Workshop on biodiversity research and information in Asia Oceania. Available from: http://www.sp2000ao.nies.go.jp/english/whats_new/year_2000/abstract.html Accessed 4 Nov 2004.

Buttler, D., Coleman, M., Critchlow, T., Fileto, R., Han, W., Liu, L., et al. (2002). Querying multiple bioinformatics data sources: Can semantic web research help? *ACM SIGMOD Record, 31*(4), 59–64.

CBD (2005). Convention on Biological Diversity (online). Available from: http://www.biodiv.org Accessed 10 Jan 2006.

GBIF (2004). Global biodiversity information facility (online). Copenhagen, Denmark. Available from: www.gbif.org Accessed 20 Nov 2005.

IABIN (2004a). Inter-American Biodiversity Information Network (online). Available from: http://www.iabin.net/english/index.shtml. Accessed 8 Nov 2005.

IMR (2006). Institute of Medical Research (online). Available from: http://www.imr.gov.my/ Accessed 8 Oct 2006.

Merican, A.F., Othman, R.Y., Ismail, N., Cheah, K.P., Mok, L., Yin, Y.K.C., & Kaur, S. (2002). Development of Malaysian Indigenous Microorganisms Online Database System. *Asia Pacific Journal of Molecular Biology and Biotechnology, 10*(1): 69–72.

Napis, S., Salleh, K.M., Itam, K., & Latiff, A. (2001). Biodiversity Databases For Malaysian Flora and Fauna: an Update. Proceedings of IWS2001 (Internet Workshop 2001), 21-23 Feb 21, 2001, National Institute of Informatics, Tokyo, Japan; organized by Internet Research Committee (IRC) of Institute of Electronics, Information and Communication Engineers (IEICE), Japan and High Quality Internet Study Group (HQI) of Information Processing Society of Japan (IPSJ)

Nilges, M., & Linge, J. P. (2002). Bioinformatics (online). Paris, France. Available from: http://www.pasteur.fr/recherche/unites/Binfs/definition/bioinformatics_definition.html Accessed 10 June 2006.

Sarinder, K. K. S., Majid, M. A., Lim, L. H. S., Ibrahim, H., & Merican, A.F. (2005). *Integrated biological database initiative (IBDI). In: Proceedings of International Conference on Biogeography and Biodiversity Wallace in Sarawak—150 years later*. Kuching, Malaysia.

Schnase, J., Cushing, J., Frame, M., Frondorf, A., Landis, E., Maier, D., & Silverschatz, A. (2003). Information technology challenges of biodiversity and ecosystems informatics. *Information Systems*, 28(4), 339–345.

Secretariat on the Convention on Biological Diversity (2005). Clearing-house mechanism (CHM) (online). Canada. Available from: http://www.biodiv.org/chm/default.aspx Accessed 10 Jan 2006.

Sivasothi, N. (2006). Raffles museum of biodiversity research (online). Department of biological sciences, The National University of Singapore. Available from: http://rmbr.nus.edu.sg/ Accessed 10 July 2006.

Wong, K. M. (2005). Rimba ilmu botanical garden (online). University Malaya, Kuala Lumpur. Available at : http://rimba.um.edu.my Accessed 15 July 2004.

Chapter 2
Preliminary Study

2.1 Introduction

In this chapter, a preliminarily study was done on an existing system, Distributed Generic Information Retrieval (DiGIR). This study was carried out in two phases described below.

2.2 First Phase Implementation

The DiGIR protocol specification is an XML-Schema document that defines the structure of messages sent to, and returned by a DiGIR provider. As illustrated in Fig. 2.1, it provides a single point of access (portal) to distributed information resources. It enables search and retrieval of structured data and makes location and technical characteristics of native resource transparent to users. DiGIR (Distributed Generic Information Retrieval) protocol was used as the communication protocol for queries between distributed databases (Fig. 2.1). As for the data sharing/exchange format, the Darwin Core (DwC) schema was used.

2.2.1 Implementation of DiGIR Using an Algae Database

The Distributed Generic Information Retrieval (DiGIR) Provider has been installed and tested using an Algae database in a server physically located at University of Malaya. The Algae database is part of the Malaysian Microbial Online Database (MIMODS) (Merican et al. 2002). The procedure of installing a DIGIR provider can be briefly outlined in a few steps.

a. Installation of PHP on the web server
b. Installation of DiGIR provider distribution
c. Editing of localconfig.php to reflect local installation choices

S. K. Dhillon and A. S. Sidhu, *Data Intensive Computing for Biodiversity*,
Data, Semantics and Cloud Computing 485, DOI: 10.1007/978-3-642-38047-1_2,

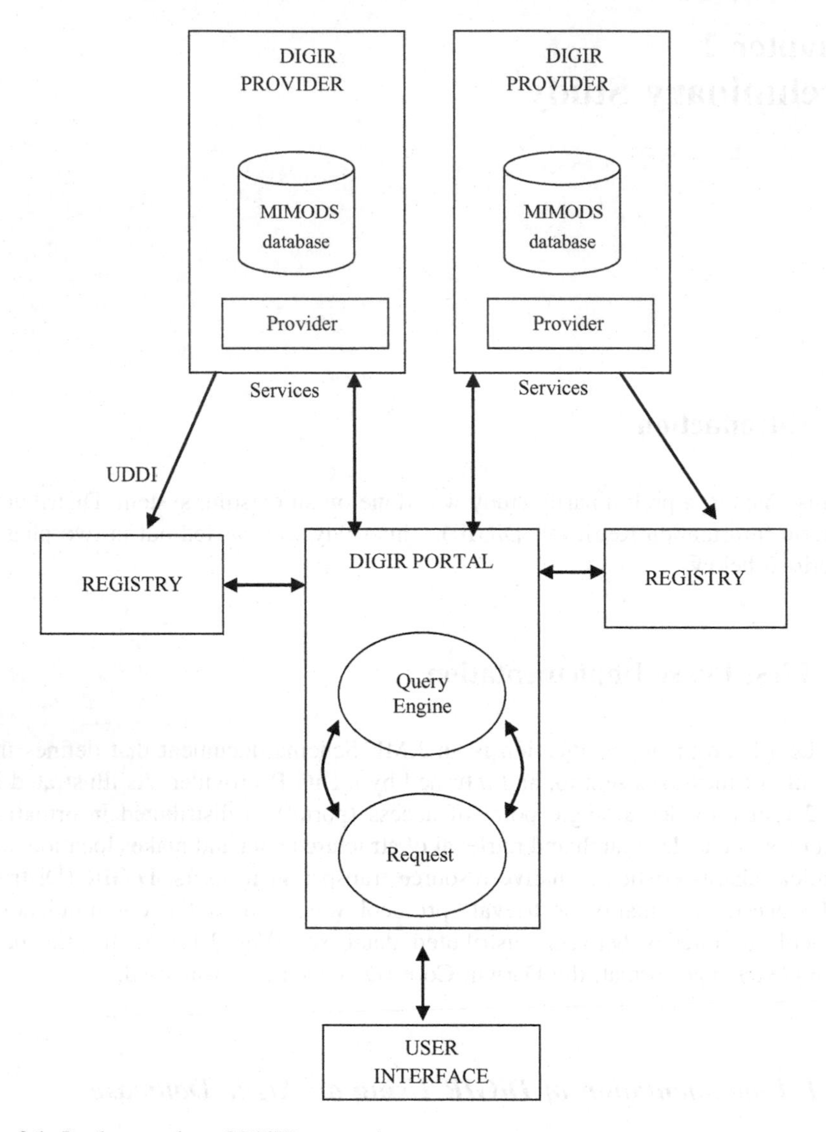

Fig. 2.1 Implementation of DiGIR protocol

d. Editing of providerMeta.xml to present appropriate metadata about the installation
e. Creating the database configuration file and saving it in the configuration folder
f. Updating resources.xml to point to the new configuration file.

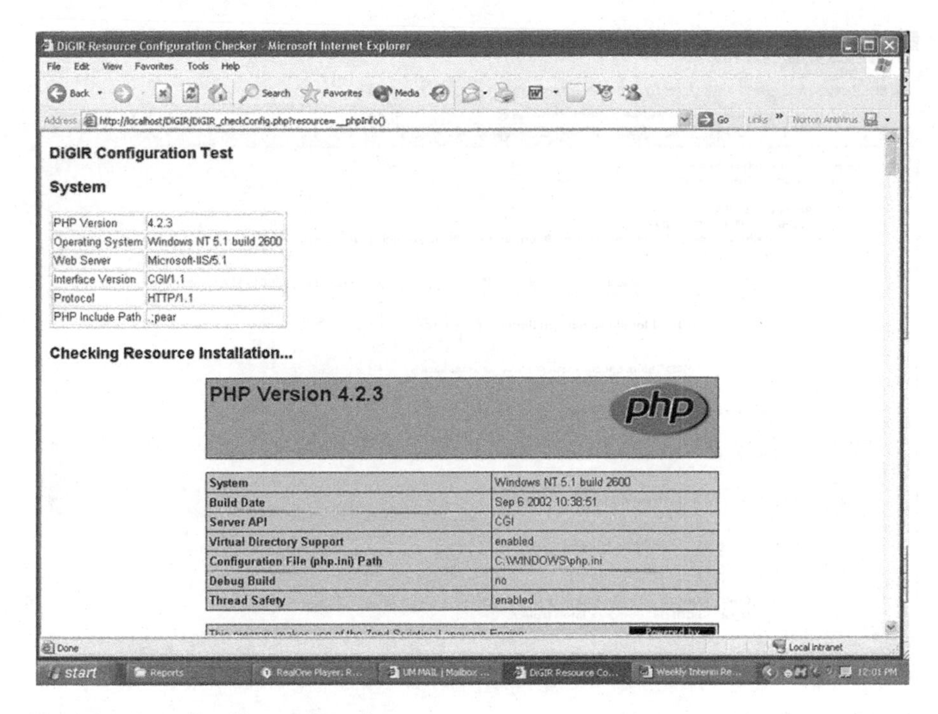

Fig. 2.2 DIGIR provider configuration

An Adodb database connectivity was used to establish a connection with the Ms Access test database. The installation showed that DiGIR Provider could run under the following specifications (Fig. 2.2).

The installation in Fig. 2.2 was based on the Bioinformatics server system configuration in University of Malaya (http://www.bioinformatik.um.edu.my). The MIMODS databases were stored in this server and a DiGIR provider is being developed to serve these databases.

A scan was done to check the Algae database contents. The scan was invoked using the "scan" operation parameter. The results of the scan in XML format are displayed in Fig. 2.3.

The results in Fig. 2.3 will be extracted by the DiGIR portal, which should be installed in a remote machine.

A simple search operation was also run to check whether it extracts the correct data from the database. This simple search looks for the string "otus %" in attribute #1 which maps to "darwin:ScientificName?". The query is displayed in the following URL:http://127.0.0.1/DiGIR/DiGIR.php?operation=search&resource=test&filter=@attr+1=1+%22otus%25%22

An XML format results were generated (Fig. 2.4).

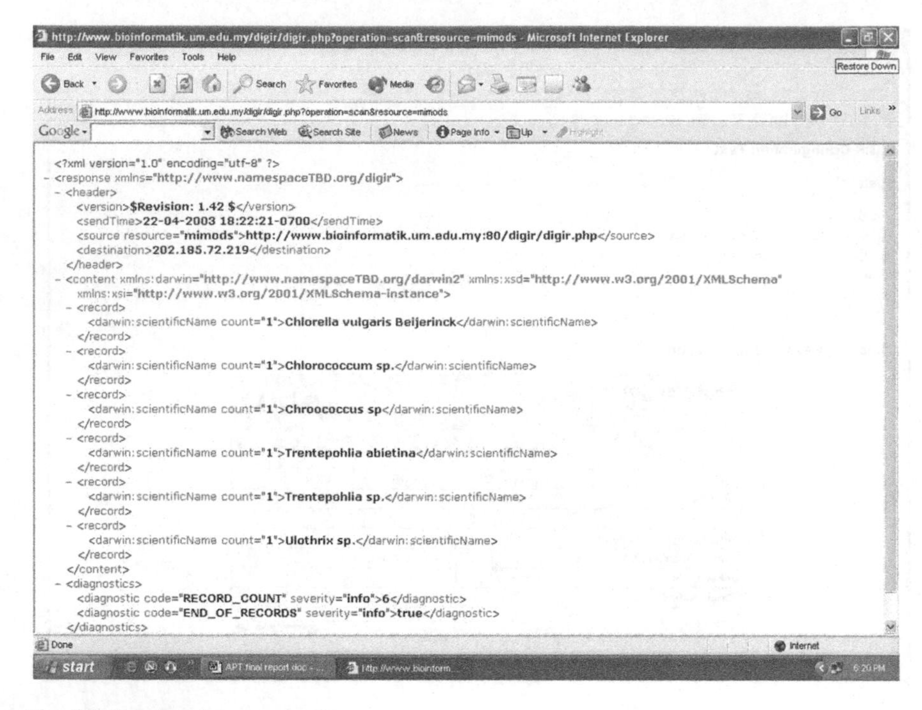

Fig. 2.3 Results of scan operation

2.2.2 Installation of a Simple Portal on a Remote Machine

There was a problem installing the DiGIR portal to search these databases using the interface. The installation of DiGIR's presentation layer was deemed to be unsuccessful in the context of this project and due to time limitation, we decided to develop our own portal (Fig. 2.5).

A simple portal was designed to search the MIMODS provider remotely. Figure 2.5 shows portal interface.

The portal in Fig. 2.5 is installed on a remote machine, which is running on a Linux platform. The portal interface was designed using Ruby-programming language.

By entering an algae family name (see Fig. 2.5), the meta search engine can extract relevant data from MIMODS local algae database.

The results of the search are displayed in a tabular form (see Fig. 2.6). The search option is currently limited to microbial family name in the algae database in MIMODS.

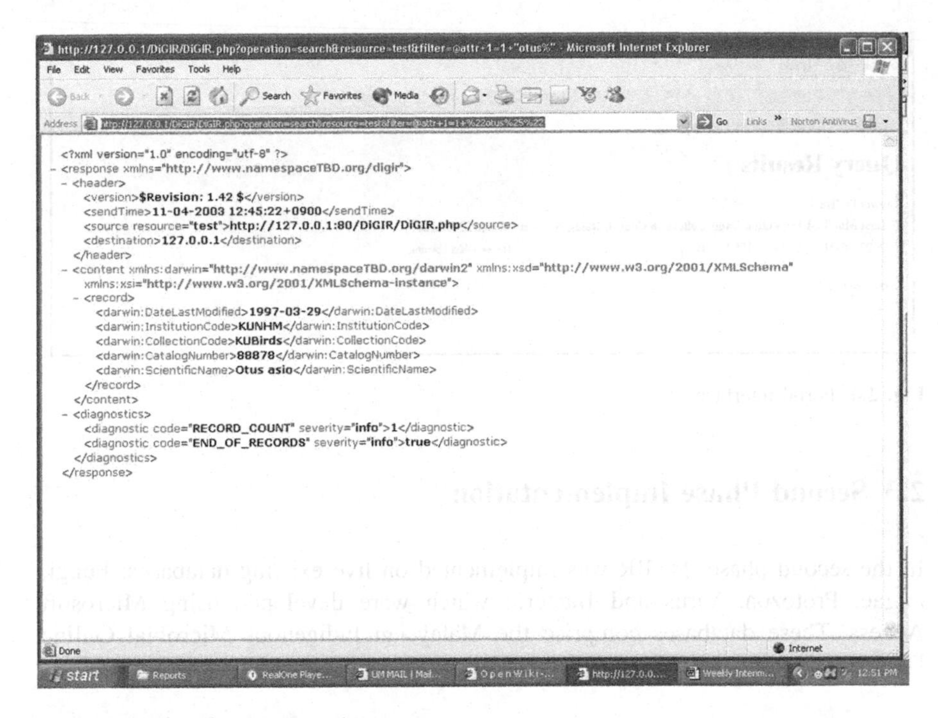

Fig. 2.4 Results of search operation

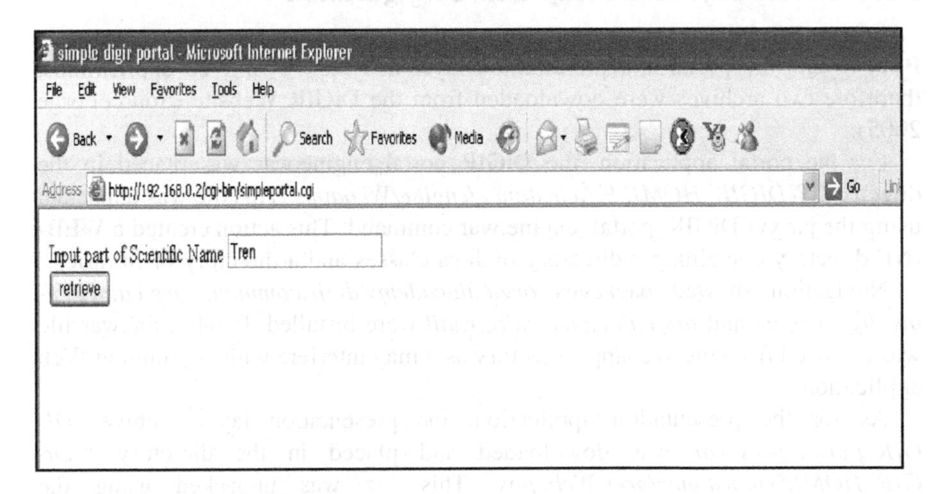

Fig. 2.5 Mainpage of DiGIR system

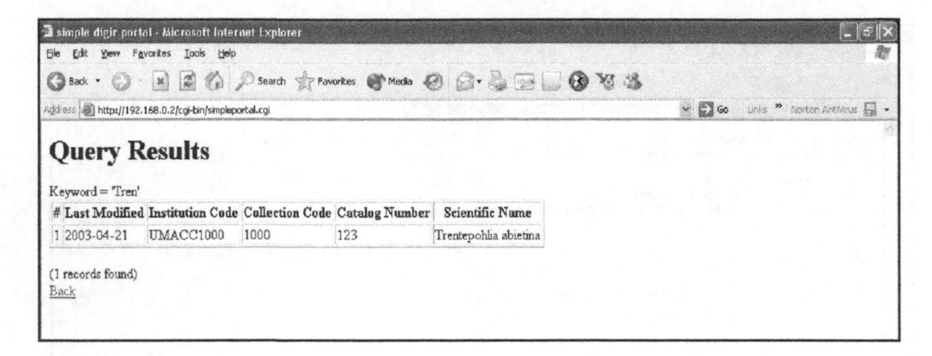

Fig. 2.6 Portal interface

2.3 Second Phase Implementation

In the second phase, DiGIR was implemented on five existing databases; Fungi, Algae, Protozoa, Virus and Bacteria which were developed using Microsoft Access. These databases comprise the Malaysian Indigenous Microbial Online Database System (MIMODS).

2.3.1 DiGIR Software Setup and Configuration

Both layers, the portal and presentation layer had their own Web applications, therefore two archives were downloaded from the DiGIR Website (SourceForge 2005).

For the portal application, the DiGIR_portal_engine.war was placed in the directory *%DIGIR_HOME %/tcinstance/engine/Webapps*. This file was unpacked using the jar xvf DiGIR_portal_engine.war command. This action created a WEB-INF directory containing a directory of Java classes and a directory of libraries.

Navigation showed packages *org.calacademy.digir.common, org.calacademy.digir.engine* and *org.calacademy.digir.util* were installed. Finally, the.war file was removed from the Webapps directory as it may interfere with the running Web application.

As for the presentation application, the presentation layer archive, *DIGIR_portal_pres.war* was downloaded and placed in the directory *%DiGIR_HOME%tcinstance/pres/Webapps*. This file was unpacked using the command *jar xvf DiGIR_Portal_pres.war*. Hence, WEB-INF directory was created containing a directory of java classes and a directory of libraries. Java packages *org.calacademy.digir.common*, *org.calacademy.digir.presentation* and *org.calacademy.digir.util* were also installed. Like the portal setup, the.war file was also removed in this layer.

Table 2.1 Portal.xml file configuration: the coding in bold is important for inserting a provider into the portal

<?xml version="1.0" encoding="UTF-8"?>
<configxmlns=http://www.calacademy.org/portal
xmlns:xsi="http://www.w3.org/2001/XMLSchema-instance"
xsi:schemaLocation="http://www.calacademy.org/portal.xsd">
<version>0.91</version>
<logfileName>/DiGIR/logs/portal_services.log</logfileName>
<providerFilter>org.calacademy.digir.engine.Darwin2ProviderFilterer</providerFilter>
<handlerPoolThreadMax>10</handlerPoolThreadMax>
<registry>
<uddiInquiryURL>http://uddi.microsoft.com/inquire</uddiInquiryURL>
<uddiServiceKey>UUID:4DFAB7E8-6387-431D-BC20-6291E99A51A8</uddiServiceKey>
</registry>
<providerCacheTimeout>0</providerCacheTimeout>
<provider>
<name>MIMODS Provider</name>
<accessPoint>http://202.185.72.215/digir/DiGIR.php**</accessPoint>**
</provider>

The engine configuration involved editing the portal.xml file located in the *%DiGIR_HOME/tcinstance/engine/Webapps/WEB-INF/classes/org/calacademy/digir/engine* directory. The configuration of the portal.xml is displayed in Table 2.1.

For the presentation layer, the presentation.xml file was configured. The configuration of the presentation.xml file is displayed in the Table 2.2.

In summary, the implementation of DiGIR comprised of the engine and presentation layer. The two most important files are essentially the portal.xml and presentation.xml file.

Table 2.2 Presentation.xml file configuration

<display>
<color>#663366</color>
<title>**DiGIR University Malaya BioWeb**</title>
<image>digir.gif</image>
<link>http://digir.sourceforge.net</link>
<intro><p>**This is a generic DiGIR for University Malaya test portal.**</p></intro>
<about><p>**This is a generic DiGIR for University Malaya testportal.**</p></about>
<footer>**This is the footer for the generic DiGIR test portal**. More information about DiGIR is available on the<a href="http://digir.sourceforge.net">DiGIR SourceForge Site</a></footer>
</display>

This file signifies the user interface. The text in bold were inserted to imply a University Malaya test portal

2.3.2 Provider Setup

The DiGIR provider hides the details of the underlying database. Therefore the provider file had to reside in the hosting machine of the database. The DiGIR.php file is the primary mechanism for invoking the DiGIR provider.

The Algae database is used here to provide an example for the provider setup. Algae database is one of the databases in Malaysian Indigenous Microbial Online Database Systems (MIMODS). This Algae database is represented using an XML file, which acts as the wrapper between the portal and the back end database. This XML wrapper is part of the provider, besides the database itself. XML file for the Algae database (digir_algae.xml) is presented in Table 2.3. Before this file was configured, the Algae database had to be altered as the DiGIR system required five fields to be present in the database for it to be eligible as a DiGIR provider. The five fields are *Date Last Modified, Institution Code, Collection Code, Catalogue Number and Scientific Name*.

Table 2.3 Digir_algae.xml file (This file contains important information regarding the database)

```
<?xml version="1.0" encoding="UTF-8"?>
<configuration>
<!–The database connection and location of database–>
<datasource
constr="Provider=Microsoft.JET.OLEDB.4.0;Data Source=
    "c:\digir\mimods\algae\db\algae.mdb"
pwd=""
database=""
encoding="utf-8"
uid=""
dbtype="ado_access"
type="SQL">
</datasource>
<table name="algae" key="species"></table>
<defaultFormats>
<format type="search"
location="http://digir/sourceforge.net/schema/conceptual/darwin/brief/2003/1.0/
    darwin2brief.xsd"/>
</defaultFormats>
<concepts xmlns:darwin="http://digir.net/schema/conceptual/darwin/2003/1.0">
<!–Concepts refers to database fields with its attributes as in the relational database–>
<concept searchable="1" returnable="1" name="darwin:DateLastModified" type="datetime"
    table="algae" zid="10" field="datelastmodified"/>
<concept searchable="1" returnable="1" name="darwin:InstitutionCode" type="text"
    table="algae" zid="10" field="institutioncode"/>
<concept searchable="1" returnable="1" name="darwin:CollectionCode" type="text"
    table="algae" zid="11" field="collectioncode"/>
```

(continued)

Table 2.3 (continued)

```
<concept searchable=“1” returnable=“1” name=“darwin:CatalogNumber” type=“text”
   table=“algae” zid=“12” field=“catalognumber”/>
<concept searchable=“1” returnable=“1” name=“darwin:ScientificName” type=“text”
   table=“algae” zid=“11” field=“scientificname”/>
<concept searchable=“1” returnable=“1” name=“darwin:Family” type=“text” table=“algae”
   zid=“1” field=“family”/>
<concept searchable=“1” returnable=“1” name=“darwin:Genus” type=“text” table=“algae”
   zid=“1” field=“genus”/>
<concept searchable=“1” returnable=“1” name=“darwin:Species” type=“text” table=“algae”
   zid=“2” field=“species”/>
</concepts>
<!–schemaLocation is the physical location of the federation schema that this resource supports.
   The value of the element is the name space of the schema.–>
<conceptualSchema schemaLocation=“http://digir.net/schema/conceptual/darwin/2003/1.0/
   darwin2.xsd”>http://digir.net/schema/conceptual/darwin/2003/1.0</conceptualSchema>
<!–integer, specifies the minimum number of characters that can appear in a query term–>
<minQueryTermLength v=“0”/>
<!–integer, specifies the maximum number of records that can be returned in a search response.
   Effectively sets the maximum page size of a response–>
<maxSearchResponseRecords v=“1000”/>
<!–integer, specifies the maximum number of records that can be returned in an inventory (scan)
   response. Effectively sets the maximum page size of a response–>
<maxInventoryResponseRecords v=“10000”/>
!–URL, points to the default record format that will be used when the request contains a blank
   record structure–>
<defaultRecordFormat v=“http://digir.sourceforge.net/schema/conceptual/darwin/brief/2003/1.0/
   darwin2brief.xsd”/>
<!–namespace+ concept, identifies the concept that will be used by default is the inventory
   request contains a blank record structure–>
<defaultInventoryConcept v=“darwin:ScientificName”/>
</configuration>
```

The provider setup completes the implementation of DiGIR. The results are presented in Sect. 2.3.4.

2.3.3 Running the DiGIR Portal

In order to see the results, the engine and presentation layers of DiGIR must be started. The steps performed are described in (a) and (b). The results of the implementation are explained in Sect. 2.3.4.

2.3.3.1 Engine Startup

The commands (1) and (2) were executed at the command prompt to run the engine;

cd %DIGIR_HOME%/tcinstance/engine/bin (1)
startup_engine (2)
Next, the URL (3) was typed in the Web browser to test whether the engine is running
http://YOUR.IP.OR.DOMAIN:8080/portal/PortalServlet?action=getProviders (3)

2.3.3.2 Presentation Startup

Following the start of the DiGIR engine, the presentation was started with the commands (4) and (5)

cd %DIGIR_HOME%/tcinstance/pres/bin (4)
startup_pres (5)
Next, the URL (6) was typed in the Web browser to test whether the presentation is running
http://YOUR.IP.OR.DOMAIN:10080/pres/PresentationServlet?action=home (6)

2.3.4 DiGIR Implementation Results

Following the startup processes described in Sect. 2.3, the DiGIR application was run to examine its results. However, the results showed unexpected outcome. The unsatisfactory results are presented in the following sub-sections.

2.3.4.1 Main Page

The DiGIR mainpage was presented on URL (7) (see Fig. 2.7)http://localhost: 10080/PresentationServlet?action=home (7)

The build query button in Fig. 2.8 takes the user to the query form (see Fig. 2.9).

2.3.4.2 Query Form

The build query button in Fig. 2.8 takes the user to the page in Fig. 2.9 which shows the query form. Here the providers and search conditions were displayed. The highlighted providers are the MIMODS databases (Malaysian Indigenous Algae, Malaysian Indigenous Bacteria and Malaysian Indigenous Fungi). The conditions of query must be selected in order to see the results.

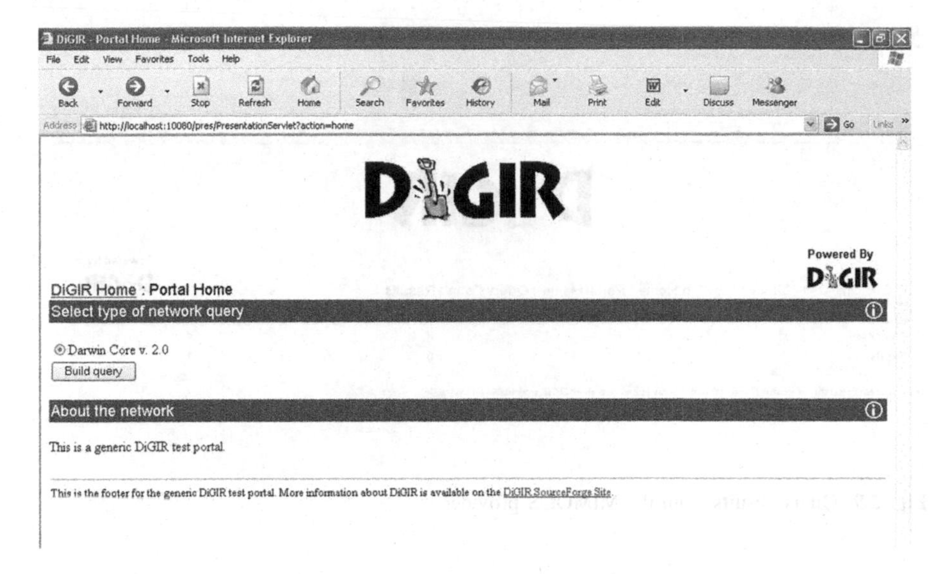

Fig. 2.7 Query results

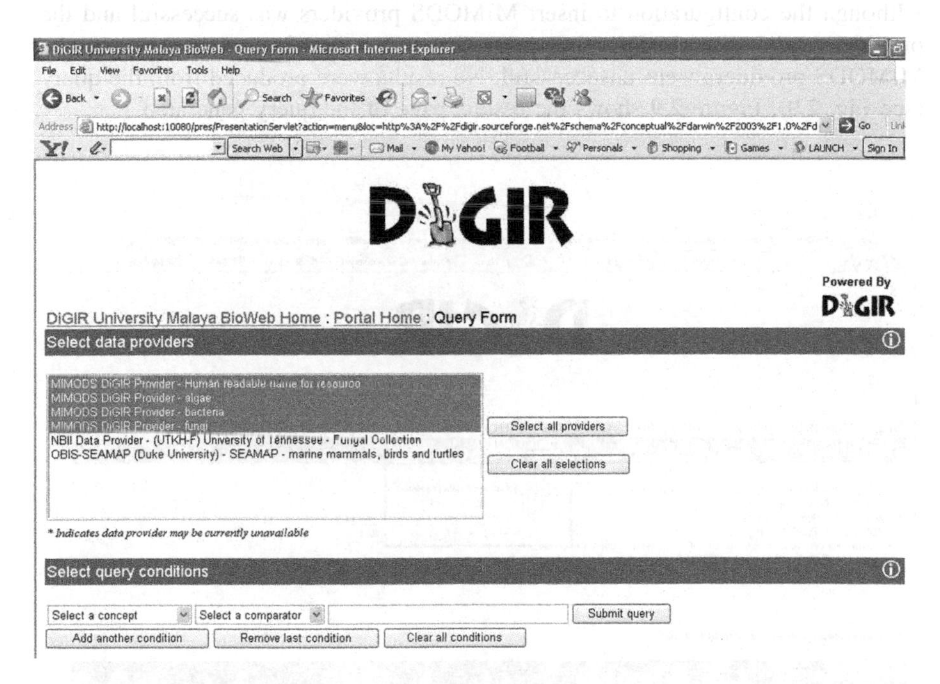

Fig. 2.8 Query form in the DiGIR system

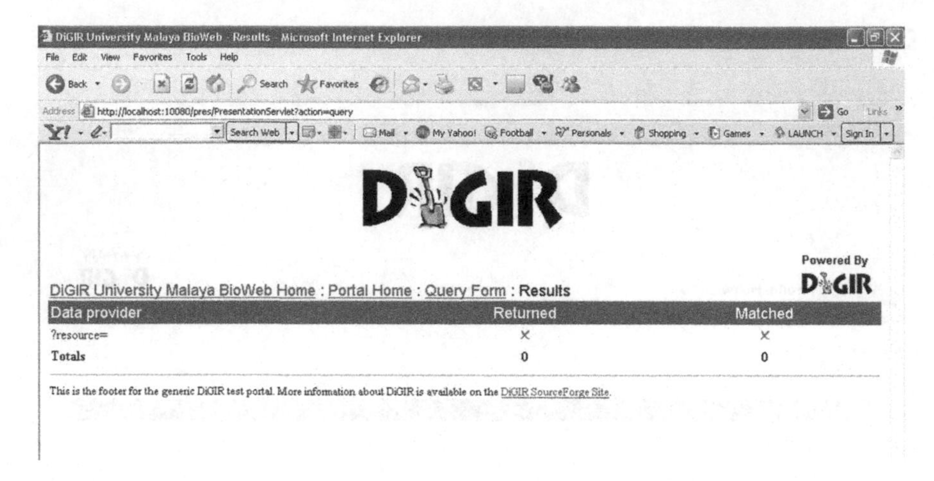

Fig. 2.9 Query results from the MIMODS providers

2.3.4.3 Query Results

Although the configuration to insert MIMODS providers was successful and the providers were displayed in the query form (see Fig. 2.9), the query on the MIMODS providers were unsuccessful. No results were produced from the query (see Fig. 2.9). Figure 2.9 shows the results page of the query generated.

Fig. 2.10 The error message indicates that no data provider was selected by user

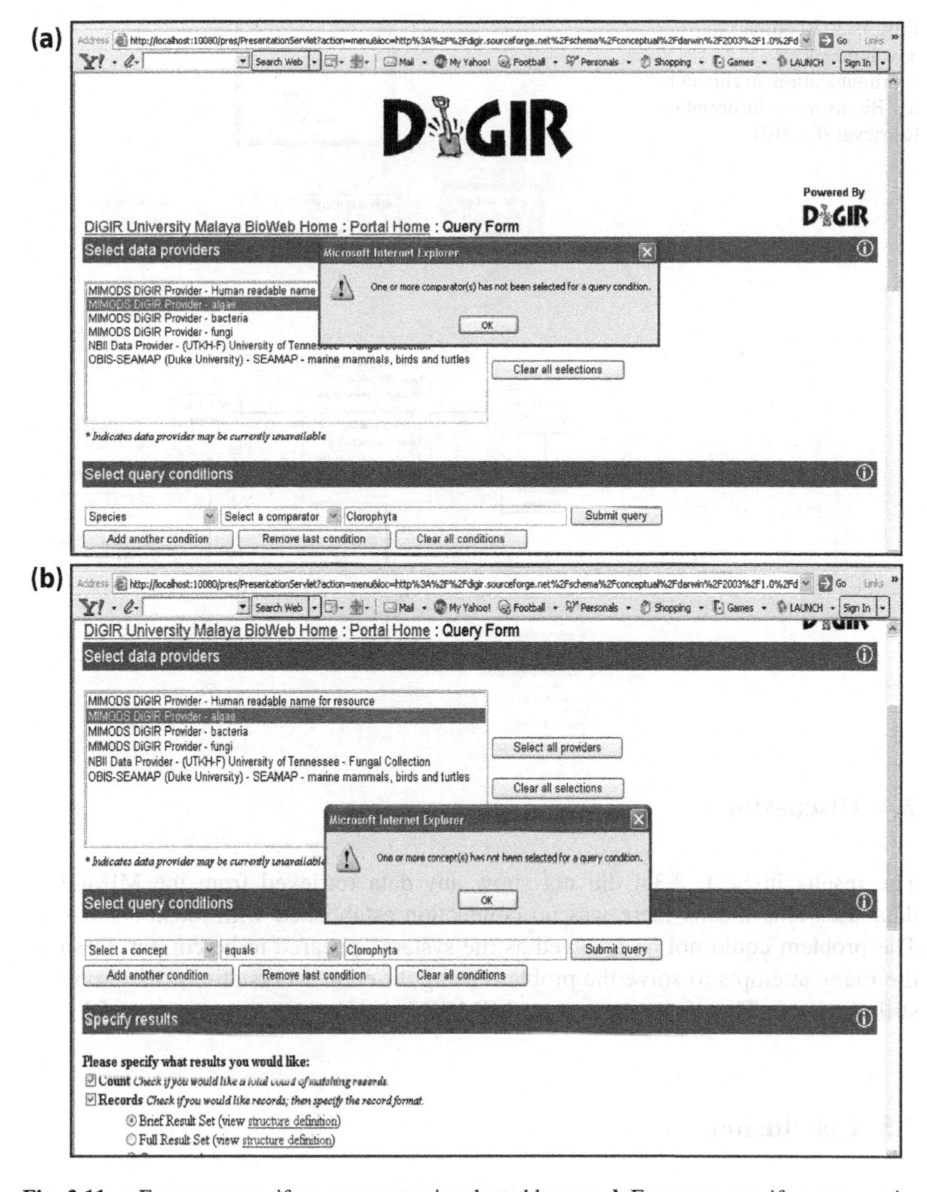

Fig. 2.11 **a** Error message if no comparator is selected by user. **b** Error message if no concept is selected by user

2.3.4.4 Other Features

The DiGIR system provides feedback to users in case of any mistake done. For example if a user does not select any provider, an error text message will appear informing the user that there was no provider selected. Figures 2.10 and 2.11 show examples of error messages in the DiGIR system.

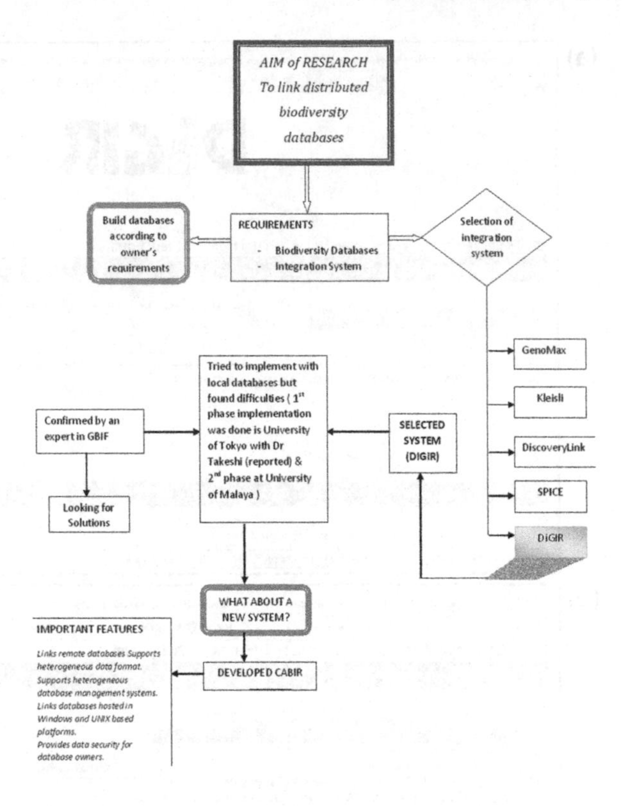

Fig. 2.12 Roadmap to development of Communication Architecture for Biodiversity Information Retrieval (CABIR)

2.4 Discussion

The results in Sect. 2.3.4 did not show any data retrieved from the MIMODS database. This means there was no connection established with those databases. The problem could not be resolved as the system appeared to be correct. Despite the many attempts to solve the problem using different workstations, the problem still persisted. Therefore, the test on MIMODS databases was not successful.

2.5 Conclusion

The failure of DiGIR implementation has lead to the development of a new system in this research. The roadmap to development of Communication Architecture for Biodiversity Information Retrieval (CABIR) is illustrated in Fig. 2.12. The system developed in this research which is named CABIR, used DiGIR as a model but it is much simpler compared to DiGIR. The results of the database integration system are presented in Chapter 6.

References

SourceForge (2005) Distributed Generic Information Retrieval (DiGIR) (online). Available from http://digir.sourceforge.net/. Accessed 20 Mar 2006.

Merican, A.F., Othman, R.Y., Ismail, N., Cheah, K.P., Mok, L., Yin, Y.K.C., & Kaur, S. (2002). Development of Malaysian Indigenous Microorganisms Online Database System. *Asia Pacific Journal of Molecular Biology and Biotechnology, 10*(1): 69–72.

Chapter 3
Literature Review

3.1 Introduction

Every research work goes through a very crucial stage, which is studying literatures related to the research. In this chapter, literatures concerning the area of research are reviewed thoroughly. Due to the multidisciplinary nature of this research, the literature review involved many different aspects which are discussed below.

Initially, a comprehensive study was carried out to understand the architecture of several existing database integration systems in biology which are GenoMax (InforMax 2001), Kleisli (Wong 2000a), DiscoveryLink (Haas et al. 2001), SPICE (Jones et al. 2000) and DiGIR (Biodiversity Research Center 2005). The strengths and weaknesses of each of these systems were also studied. Then, the underlying components and technologies which are essential to a database integration system were researched, such as databases, Database Management Systems and data formats. Finally, a study was done in the technical areas of integration approach, interface design, data integration with the Web, programming and scripting languages, query language, database connectivity and Web services. Based on these studies, a novel system for database integration was proposed.

3.2 Existing Database Integration System

The systems studied were GenoMax, Kleisli, DiscoveryLink, SPICE and DiGIR. For each of these systems, the strength and weaknesses are discussed (see below).

3.2.1 GenoMax

GenoMax is an enterprise-level integration of bioinformatics tools and data sources developed by InforMax (2001). It is a good illustration of an amalgamation of a few point solutions, including a sequence analysis module and a gene

S. K. Dhillon and A. S. Sidhu, *Data Intensive Computing for Biodiversity*,
Data, Semantics and Cloud Computing 485, DOI: 10.1007/978-3-642-38047-1_3,

expression module, developed on top of a data warehouse of fixed design (3rd Millennium Inc. 2002). The warehouse is an ORACLE database designed to hold sequence data, gene expression data, 3D protein structures, and protein–protein interaction information. Load routines are built-in for standard data sources such as GenBank (nucleotide sequences database; http://www.ncbi.nlm.nih.gov/Genbank/) and SWISS-PROT (protein knowledgebase; http://www.ebi.ac.uk/swissprot/). A special scripting language of limited expressive power is also supported for building analytical pipelines.

Its strengths are twofold (Wong 2002). Firstly, each of GenoMax's component point-solution modules is a very well designed application for a specific purpose. For example, its gene expression module provides Self-Organizing Map (SOM) clustering, Principal Component Analysis (PCA) and so forth on microarray data via simple-to-use graphical user interfaces. Secondly, these components are tightly integrated via a specially designed data warehouse. Its weakness is its tight point-solution-like application integration. GenoMax covers less data types and products than systems such as Kleisli and DiscoveryLink (Sect. 3.2.3). For example, these latter systems can easily incorporate chemical data which are beyond the current data warehouse design of GenoMax. In addition, GenoMax's scripting language is not designed for large-scale database style manipulations, hence, these types of *ad hoc* queries are neither straightforward nor optimized in GenoMax. There are also difficulties in adding new kinds of data sources and analysis tools Fig. 3.1.

3.2.2 Kleisli

Kleisli (Wong 2000a) is marketed by geneticXchange Inc. of Menlo Park, United States. It is one of the earliest systems that have been successfully applied to some of the earliest data integration problem in the Human Genome Project (HGMIS 2005).

The approach taken by the Kleisli system is illustrated in Fig. 3.1. It is positioned as a mediator system encompassing a nested relational data model, a high-level query language, and a powerful query optimizer. It runs on top of a large number of light-weight wrappers for accessing various data sources. There are also a number of application programming interfaces that allow Kleisli to be accessed in an ODBC- or JDBC-like fashion in various programming languages for a various applications. The Kleisli system is highly extensible. It can be used to support several different high-level query languages by replacing its high-level query language module. Currently, Kleisli supports a "comprehension syntax"—based language called CPL (Wong 2000a) and a "nested relationalized" version of SQL called sSQL. The Kleisli system can also be used to support many different types of external data sources by adding new wrappers, which forward Kleisli's requests to these sources and translate their replies into Kleisli's exchange format. These wrappers are light weight and new wrappers are easily developed and inserted into the Kleisli system. The optimizer of the Kleisli system can also be customized by different rules and strategies (Wong 2000a). Kleisli does not have

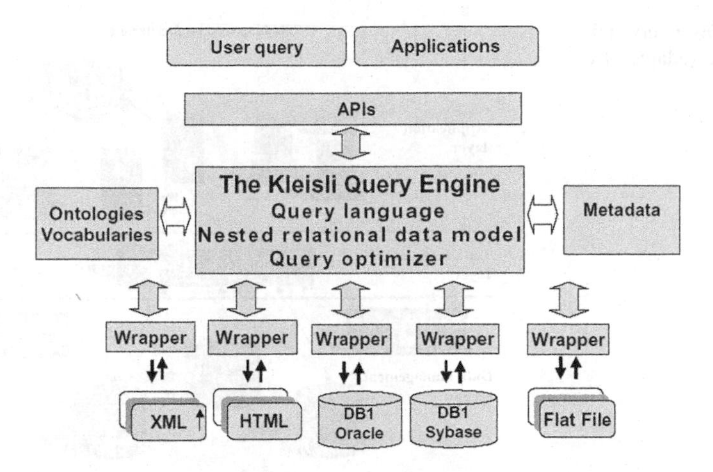

Fig. 3.1 Kleisli architecture (adapted from Wong 2002)

its own native database management system. Instead, Kleisli has the ability to turn many kinds of database systems into an updatable store conforming to its nested relational data model. In particular, Kleisli can use flat relational database management systems such as Sybase, Oracle and MySQL to be its updatable store. It can even use all of these systems simultaneously. It is also worth noting that Kleisli stores nested relations into flat relational database management systems using an encoding scheme that does not require these nested relations to be fragmented over several tables (Wong 2000a).

Kleisli possesses the following strengths. (1) it does not require data schemas to be available. It has a nested relational data model and a data exchange format that external databases and software systems can be easily translated into. It shields existing queries, via a type inference mechanism, from certain kinds of structural changes in the external data sources, (2) it has the ability to store, update, and manage complex nested data. It has a good query optimizer, (3) it is equipped with two application programming interfaces so that it can be accessed in a JDBC-like manner from Perl and Java (Wong 2000b). However, according to Wong (2000b), Kleisli shares a common weakness with DiscoveryLink (Sect. 3.2.3). Even though CPL and sSQL are both high-level query languages and protect the user from many low level details such as communication protocols, memory management and thread scheduling, the programming of queries is probably still beyond the expertise of an average biologist.

3.2.3 Discovery Link

DiscoveryLink (Haas et al. 2001) is an IBM product and, in principle, a general data integration system for biomedical data (see Fig. 3.2). This relational data model dictates the way a DiscoveryLink user views the underlying data, results

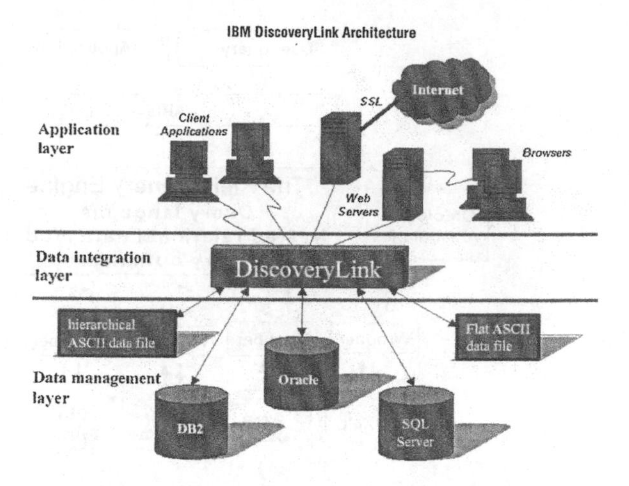

Fig. 3.2 DiscoveryLink architecture (adapted from Hall 2002)

and queries the database. The relational data model is the *de facto* data model of most commercial database management systems, including the IBM's DB2 database management system upon which DiscoveryLink is based. As a result, DiscoveryLink comes with a high-level query language, SQL that is a standard feature of most Database Management Systems.

This gives DiscoveryLink several advantages. Firstly, not only can a user easily express SQL queries that go across multiple data sources, but she can also perform further manipulations on the results. Secondly, not only are the SQL queries more powerful and expressive, the SQL queries are also automatically optimized by DB2. The use of query optimization allows a user to concentrate on getting the query right without worrying about getting it fast (Wong 2002).

However, DiscoveryLink has limitations as a tool for creating and managing data warehouses for biological data (Wong 2002). For example: (1) Limited number of "wrappers" which is a hindrance for this system to expand (Hall 2002) (2) The system is too generic in the sense that the design is not specific enough to the life sciences domain to earn that status of a necessary application for scientists (Hall 2002) (3) The system is tied to the relational data model. This implies every piece of data that it handles must be a table of atomic objects like strings and numbers. Unfortunately, most of the data sources in biology are not that simple and are deeply nested. Therefore, there is severe impedance mismatch between these sources and DiscoveryLink. Consequently, it is not straightforward to add new data sources or analysis tools into the system. For example, to put the SWISS-PROT database into a relational database in the third normal form would require breaking every SWISS-PROT record into nearly 30 pieces in a normalization process. Such a normalization process requires a certain amount of skill. Similarly, to query the normalized data in DiscoveryLink requires some mental and performance overhead, as users need to figure out which part of SWISS-PROT has been assigned to which of the 30 pieces and they need to join some of the pieces back again (Wong 2002). (4) DiscoveryLink supports only wrappers written in

C++, which is not the most suitable programming language for writing wrappers. In short, it is difficult to extend DiscoveryLink with new sources (Wong 2002). (5) DiscoveryLink does not store nested objects in a natural way and is very limited in its capability for handling long documents (Wong 2002).

According to Wong (2002), in spite of the weaknesses discussed above, in theory, DiscoveryLink has greater generality than specialized application integration like GenoMax. Unfortunately, this greater generality is achieved at the price of requiring that SQL be used for expressing queries. While writing queries in SQL is generally simpler than writing in Perl scripting language, it is probably still beyond the skill of an average biologist. This is a disadvantage in comparison to GenoMax, which has good user interface for a biologist to build simpler queries.

3.2.4 SPICE

The SPICE (Species 2000 Interoperability Co-ordination environment) project is allied to Species2000 with the specific brief of researching the Computer Science challenges associated with building a scalable infrastructure for Species2000 (Jones et al. 2000). Species 2000 is a "federation" of database organizations working closely with users, taxonomists and sponsoring agencies. The goal of the Species 2000 project is to create a validated checklist of the entire world's species (plants, animals, fungi and microbes). The program in partnership with the Integrated Taxonomic Information system (ITIS) of North America currently produces the Catalogue of Life—an Annual Checklist available on the Species 2000 Web site (Species2000 2000). This is used by the Global Biodiversity Information facility (GBIF) as the taxonomic backbone to its Web portal.

The SPICE system can be divided into two parts which are the *common access system* (CAS), which acts as a World-Wide Web server (brokering requests from users) and co-ordinates the formulation of suitable queries to the *global species databases* (GSD) and the assembly of results, and *wrappers*, which wrap the GSDs into a common data model and materialize on the CORBA (Siegel 1998) 'bus' architecture so that the CAS can query the GSDs (Jones et al. 2000).

The general architecture chosen for SPICE is illustrated in Fig. 3.3. It is COBRA-based system which can be decomposed into discrete, maintainable objects. The objects are distributed optimally and the implementation language independence and platform independence of COBRA ensure that SPICE can interoperate effectively across all databases of interest (Wong 2000a).

However, SPICE does not do all the work necessary to bring each individual database into the system, but to facilitate the building of suitable wrappers. The interface of SPICE allows for searching through scientific names and common names only. Besides the above, SPICE imposes a set of requirements on the kinds of data model that the individual databases can have. In particular, there is a firm definition of the data that must be available to SPICE, and so it is natural to specify

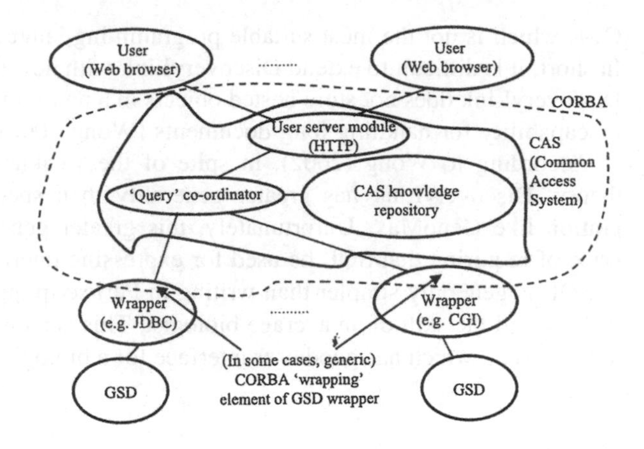

Fig. 3.3 SPICE architecture (adapted from Jones et al. 2000)

a common data model requiring all wrappers to map between the databases and that common data model (Wong 2000a).

3.2.5 Distributed Generic Information Retrieval

DiGIR is a protocol and a set of tools for linking a community of independent databases into a single, searchable "virtual" collection. The DiGIR protocol was developed by Biodiversity Research Center (BRC) Informatics in collaboration with the Museum of Vertebrate Zoology at UC Berkeley and the California Academy of Sciences. DiGIR is currently a public open source project with an international team of contributors, including Centro de Referência em Informação Ambiental (CRIA), Brazil (Biodiversity Research Center 2005).

DiGIR provides a uniform interface for managing XML-based search requests to a community of dissimilar data sources. Each institution in the DiGIR community implements an interface application called a DiGIR provider. The DiGIR provider hides the details of the underlying database and presents a uniform "virtual" view of the data to the network (Biodiversity Research Center 2005).

When a user on the network types a search request, the client application sends a query in the virtual format. Each DiGIR provider translates the search request into an equivalent request compatible with the structure of the local database. When the DiGIR provider receives a response from the local database, it translates the response to match the virtual structure and sends the result to the client (Biodiversity Research Center 2005) (see Fig. 3.4).

The main motivation of DiGIR was to unify diverse networks in a single technology. Part of its strategy was the use of open standards and protocols which are HTTP (Hypertext Transfer Protocol), XML and UDDI (Universal Description, Discovery and Integration).

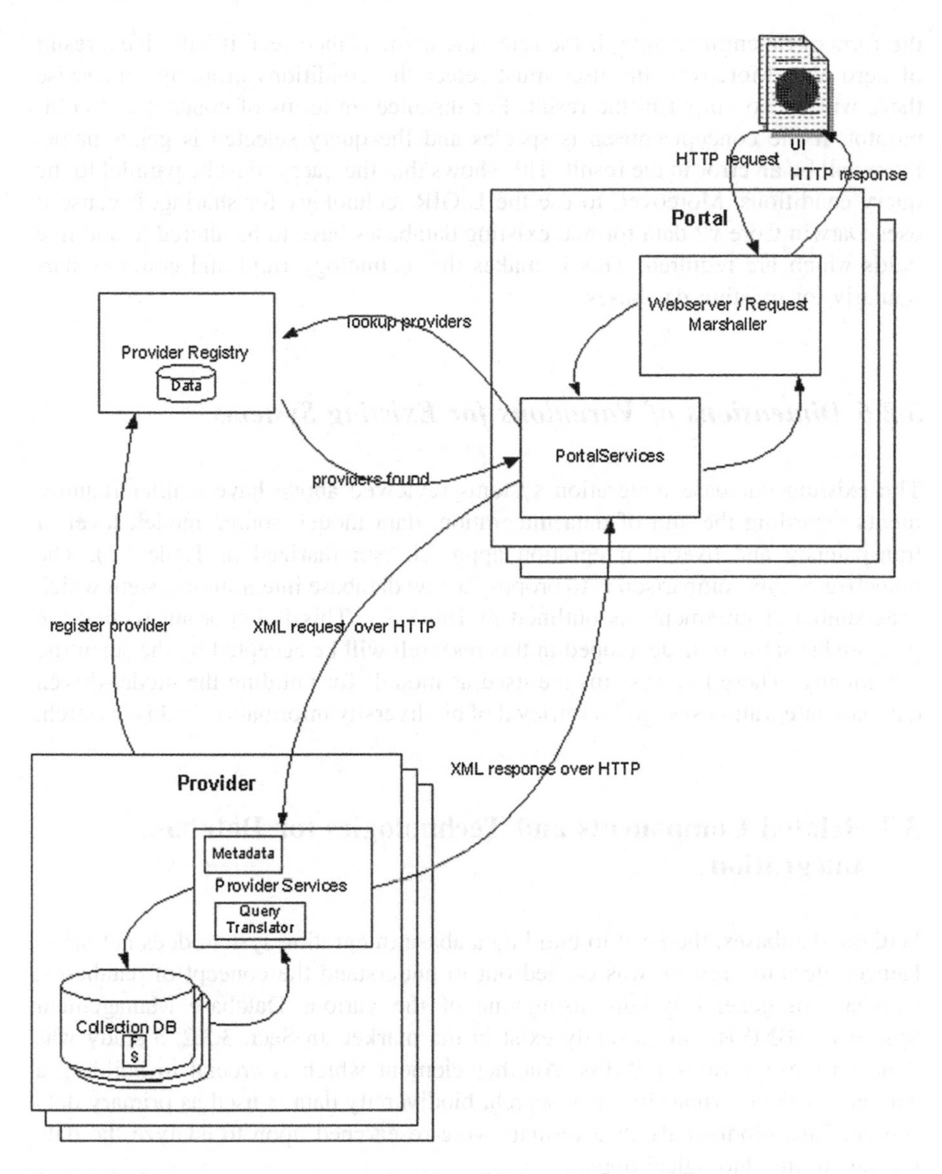

Fig. 3.4 DiGIR architecture (adapted from http://digir.sourceforge.net/)

DiGIR uses the Darwin Core data format which is a standard set of concepts describing a natural history specimen (see Sect. 3.3.3.3 for details).

DiGIR is useful in terms of its features. The user interface is easy to understand. There are options to select data providers, select query conditions, specify results and specify timeout. There is an option for the user to save the data retrieved. The system is also properly guided. For example, there is an error message if user accidentally makes a mistake. However, the search engine only accepts query in the form of Scientific Name. If the scientific name is incorrect, it will give a result of zero hits. Moreover, the user must select the conditions properly; otherwise there will be no output in the result. For instance, in terms of concept and comparator, if the concept chosen is species and the query selected is genus name, there will be an error in the result. This shows that the query must be parallel to the query conditions. Moreover, to use the DiGIR technology for sharing, because it uses Darwin Core v2 data format, existing databases have to be altered to add five fields which are required. This is makes the technology rigid and complex particularly for existing databases.

3.2.6 Dimensions of Variations for Existing Systems

The existing database integration systems reviewed above have similar requirements regarding the aim of data integration, data model, source model, level of transparency and overall integration approach (summarized in Table 3.1). The objective of this comparison is to propose a new database integration system which uses similar requirements as outlined in Table 3.1. This is important so that the proposed system to be developed in this research will be accepted by the scientific community. These five systems are used as models for building the model-driven database integration system for retrieval of biodiversity information in this research.

3.3 Related Components and Technologies for Database Integration

Without databases, the need to build a database integration system does not arise. Hence, literature review was carried out to understand the concept of databases. Database in general is built using one of the various Database Management Systems (DBMS)s that currently exist in the market. In Sect. 3.3.2, a study was done on these various DBMSs. Another element which is crucial to building a database is data format. In this research, biodiversity data is used as primary data source, thus, biodiversity data formats were researched upon to analyze the data formats in this biological domain.

3.3.1 Databases

A database can be described as a collection of data managed by a DBMS (Elmasri and Shamkant 1994). The central concept of a database is that of a collection of records, or pieces of knowledge. Typically, for a given database, there is a

Table 3.1 Dimensions of variations for the existing systems described in Sect. 3.2.6

	Aim of integration	Data model	Source model	User model	Level of transparency	Market status	Overall integration approach
GenoMax	Data mining	Structured static data	Mostly complementary	Expertise in software functionality, data mining tools, life science informatics analysis approaches, collaboration network, and other aspects of the user interface	Sources specified by head database	Commercial	Warehouse based
Kleisli	Query-oriented	Semi-structured, object-oriented	Mostly complementary	Expertise in query language	Sources specified by user	Commercial	Mediator-based
DiscoveryLink	Query-oriented middleware	Structured, object-relational	Mostly complementary, some overlap	Expertise in query language	Sources selected by system	Commercial	Mediator-based
SPICE	Query-oriented middleware	Structured, object-relational	Mostly complementary	Expertise in query language	Sources specified by user	Non-Commercial	Mediator-based
DiGIR	Query-Oriented	Structured, object-relational	Mostly complementary	Expertise in query language	Sources specified by user	Non-Commercial	Mediator-based

structural description of the type of facts held in that database: this description is known as a schema. The schema describes the objects that are represented in the database, and the relationships among them. There are a number of different ways of organizing a schema, that is, of modeling the database structure: these are known as database models (or data models). The model in most common use today is the relational model, which represents all information in the form of multiple related tables each consisting of rows and columns. This model represents relationships by the use of values common to more than one table. Other models such as the hierarchical model and the network model use a more explicit representation of relationships.

Many biodiversity databases were set up in the early 1980s, at that time when the Internet was not widely used, and Database Management Systems (DBMS) by themselves required advanced technical skills. Data have made available by proprietary methods, then later via static Web pages, and when flat files grew too big, server side scripts like Common Gateway Interface (CGI) scripts were used for searching and retrieving data from flat files. For data exchange, proprietary flat file formats were usually used, and several plain text flat file formats evolved. These flat files are structured by using letter codes at the beginning of each line of paragraph. These days, the number of databases that are implemented as flat files are decreasing and many databases has moved from their old flat file representations to DBMS (Jacob 2004).

Only a few years ago a considerable number of biodiversity databases were based on proprietary flat file solutions, currently most of the biodiversity databases are implemented on Database Management Systems (DBMS). Relational databases were first introduced in 1970 (Xia et al. 2002). Since then, a strong theoretical underpinning and many different Relational Database Management Systems (RDBMS) have been developed.

3.3.2 Database Integration Database Management Systems and Relational Database Management Systems

A Database Management System (DBMS) is software that defines a database, stores the data, supports a query language, produces reports and creates data entry screens.

Some of the most challenging problems in building applications arise in storing and retrieving data. Problems include conserving space, retrieving data rapidly, sharing data with multiple users at the same time, and providing backup and recovery of the data. Initially programmers had to solve these problems for every application they created. Today, the DBMS already provides some of the best solutions to these problems. Making the database the foundation of an application means that you get all of the powerful features and security without much additional work (Post 2005)

Modern database can be described as a collection of data managed by a DBMS (Elmasri and Shamkant 1994). The primary benefits provided by a DBMS are (Post 2005):-

- Minimal data redundancy
- Data consistency
- Integration of data
- Sharing of data
- Enforcement of standards
- Ease of application development
- Uniform security, privacy, and integrity
- Data independence.

The DBMS stores data efficiently. Moreover the data can be retrieved rapidly to answer any query. Although these two goals seem obvious, they can be very challenging to handle if you have to write programs from scratch every time.

The DBMS also has systems to maintain data consistency with minimal effort. Most systems enable the user to create basic rules when user defines the data. For example, Scientific Name should be taken from Species and Genus. These rules are enforced for every form, user, or program that accesses the data. With traditional programs such as file based management systems, everyone will have to follow the same rules. Additionally, these rules would be stored in hundreds or thousands of separate programs, hence, making them hard to find and hard to modify.

The DBMS, particularly the query language, makes it easy to integrate data. For example, one application might collect data on taxonomy. Another application might collect data on locality of species. If programmers created separate programs and independent files to store this data, combining the data would be difficult. In contrast, with a DBMS any data in the database can be easily retrieved, combined, and compared using the query system (Post 2005). Table 3.2 summarizes the current Database Management Systems to give an overview of their features.

RDBMS is a type of Database Management System (DBMS) that stores data in the form of related tables. Relational databases are powerful because they require few assumptions about how data is related or how it will be extracted from the database. As a result, the same database can be viewed in many different ways. An important feature of relational systems is that a single database can be spread across several tables.

3.3.3 Biodiversity Data Formats

Real time data integration from distributed sources requires not only the technology but standards and protocols are well.

Table 3.2 Comparison of database management systems (DBMS)

DBMS	Advantages	Disadvantages	Portability	Flexibility	Ease of use	Security	Accessibility
Microsoft access	Very powerful set of features Works with SQL Can support online databases Free with office 2003 for windows Includes charting Can be administered from graphical user interface	Lacks database triggers and stored procedures Only suitable for small applications	Windows	Flexibility and control of Microsoft windows™ APIs while shielding many of the frustrations associated with using high- and low-level language development environments	Easy to install and administer	Protects data using user-level security	Visual Basic, ASP
mySQL	No license cost Rich feature set Performs well on majority queries Large support base for access from many different languages Fast and can handle heavy load	Lacks database triggers and stored procedures No graphical user interface	Unix, Windows	Table types can be chosen	Easy to install and administer Requires little maintenance	Restrict users rights from an entire database down to the column level	C, C++, Perl, ASP, PHP, Java
DB2	Flexible and Cost-effective Can be administered from command line and graphical user interface Supports multiple triggers	Not simple therefore takes longer to learn Still costly although cheaper than Oracle	Unix Windows	DB2 UDB express edition is based on open industry standards, portable across industry pervasive platforms, such as Microsoft® Windows® or Linux®	Combines the power, function, and reliability of an open standards-based database server with simplicity in packaging, installation and deployment at a minimal investment cost	DB2 universal database uses a combination of external security services and internal access control mechanisms to perform this vital task. In most cases, three different levels of security are employed	Most third generation languages

(continued)

Table 3.2 (continued)

DBMS	Advantages	Disadvantages	Portability	Flexibility	Ease of use	Security	Accessibility
Oracle	Can store and execute stored procedures and functions within itself Can be administered from command line and graphical user interface Supports multiple triggers	Costly Need experienced database administrator	Unix Windows	Easily adaptable to various operating systems It offers pre-defined workflows and form-based tools	Easy to install and implement	There will be security holes if mis-configured	Most third generation languages
FileMaker	Easy to use Can build multiple tables into one document Scalable Can be administered from graphical user interface	Not quite as powerful as other databases (although it has become more powerful in recent years) Not a SQL database Does not support online databases well Plugins needed for charts and other functions	Macintosh Windows	Database engine integrated with screen layouts API for integration with 3rd party tools	Easy to use Suitable for even novice database administrators	Passwords are not inherently tied to individual users; instead, passwords link up more with "groups" that the database designer can define	Java, ASP

Traditionally, DBMS-based approaches for database integration demand the development of a target database schema to store and/or access the data as well as to provide integrated access. Biodiversity data can be really huge and complex. Without standardization, data sharing can become a daunting task (Philippi 2004).

According to (Xia et al. 2002), Standardisation can mean two things which are; (1) a standard terminology and nomenclature and (2) standard format for data submission and storage, exchange, and query. The need for standardization of biological terminology has its importance. For example, it is not uncommon that there are multiple names for one gene in different databases (Anthony and Leslie 2002).

Many standardisations have been developed or proposed in biology, such as standardisation of Gene Nomenclature (Povey et al. 2001), Biological Data Working Group (FGDC 2000), International Working Group on Taxonomic Databases (IWGTD 2005) and Inter-American Biodiversity Information Network (IABIN 2004b).

According to the Species Analyst (Vieglais 2003), natural history collections and observation data sets represent sets of observations, with each record detailing the observation of an organism, ideally at a specific geo-temporal location. In the case of collections, the observation is permanent in that the organism was collected from the field and preserved in a curated collection intended to last indefinitely. Collected specimens can be prepared in various ways, and several preparations from a single organism are not unusual (skin, skeleton, and perhaps microscope slides), thus there may be several records for a single organism, each representing the organism prepared using different techniques, but all records referring to a single observation event. Conversely, some collection records may represent a collection object that contains many organisms. Observation data sets catalog the observation of an organism, also at a specific geo-temporal location, but in this case the organism observed is not collected, and hence the observation record is the only information recorded about the organism. In both cases a taxonomic identification of the organism is attempted, with obvious consequences for accuracy of identification (a specimen available for identification to several experts compared with a potentially fleeting glimpse of an organism in the field).

Each specimen in a collection is irreplaceable, and many may provide insight about the previous geographic distribution of taxonomic units, and how this may have changed over time. Every specimen in a collection is identified by a tag, which may be hand written. Specimens may be collected in the field under somewhat extreme conditions, with sometimes vague descriptions of the location from which the specimen was collected. Some natural history collections are quite old, extending back some 300 years or so. Many collections are only partially computerized. For those collections that are wholly or partially computerized, there are no standards for the database content, schema, structure or type. Nevertheless, there is a commonality in the content of almost all collection and observation databases which may be exploited to perform ordered search and retrieval from these diverse data sets.

The guidelines providing commonality for biodiversity data are developed by groups of researchers all over the world. In this section, some of the most commonly used guidelines or data format are discussed. These standards are used for retrieving biodiversity database content regardless of the format used.

3.3.3.1 ABCD Schema

The Access to Biological Collections Data (ABCD) Schema is the product of a joint Taxonomic Data Working Group (TDWG) and The Committee on Data for Science and Technology (CODATA) initiative to develop a standard for distributed data retrieval from specimen collection databases. The schema supports data exchange for all kingdoms and for both specimen and observation records. The ABDC Schema is a GBIF (Global Biodiversity Information Facility) approved data standard that incorporates Darwin Core (DwC) elements (Limi et al. 2005) (see Appendix B).

The ABCD Schema attempts to be comprehensive and highly structured, supporting data from a wide variety of databases. It is compatible with several existing data standards. Parallel structures exist so that either (or both) atomized data and free-text can be accommodated. Version 1.2 is currently in use with the GBIF and BioCASE (Biological Collection Access Service for Europe) networks (Berendsohn 2006).

Apart from the GBIF and BioCASE networks, the potential for the application of ABCD extends to internal networks, or in-house legacy data access (e.g. datasets from external sources that shall not be converted and integrated into an institution's own data, but kept separately, though easily accessible). By defining relations between terms, ABCD is a step towards ontology for biological collections (Berendsohn 2006).

3.3.3.2 Darwin Core

More recently, distributed database networks of biological collections have taken a different approach. The Darwin Core (DwC) defines a simpler set of fields common to all taxonomic groups. The result is a simple XML schema with a small number of elements covering basic and essential information (Canhos et al. 2004).

According to Inter-American Biodiversity Information Network (IABIN 2004a), the DwC attempts to provide guidelines for addressing the commonality regardless of the underlying mechanism for storing the record content.

Basically, the DwC is a metadata profile describing the minimum set of standards for search and retrieval of natural history collections and observation databases. It includes only the core data elements that are likely to be available for the vast majority of specimen and observation records. This standard is utilized within both the Species Analyst and World Biodiversity Information Network (REMIB) networks, among others. DwC is also a Global Biodiversity Information Facility

(GBIF 2004) approved data standard for specimen collections and observations data.

According to Limi et al. (2005), the DwC is a specification of data concepts and structure intended to support the retrieval and integration of primary data that documents the occurrence of organisms in space and time and the occurrence of organisms in biological collections. Information resources falling within this scope include databases about natural history collections, living collections (i.e., zoological and botanical gardens), germplasm and genetic resource collections, and data sets produced from biodiversity survey and monitoring programs. These data resources support a wide variety of purposes and consequently they have different structures, but all can contribute to documenting the distributions of organisms. A secondary function of the DwC is to enable a user to discover the contents of biological collections. Since the biological collections are diverse, the DwC supports the search and retrieval of descriptive information in relatively simple ways.

The following design goals have shaped the DwC (TDWG 2006):-

- The DwC must support the combination of data from multiple sources into a structured data set.
- The structure and content should minimize the additional effort required to support further manipulation and analysis of the data.
- Every record must be identifiable by a required set of data concepts (elements or attributes), within the context of the returned data set (result of a search) and ideally within the context of all resources made available through a system, portal, or registry.
- The DwC should support queries that use data values in a semantic context. For example, a query for Collector = "Smith" should return different records than a query for IdentifiedBy = "Smith".
- The DwC should support basic data types so that appropriate comparison operators can be used in queries; for example, ElevationInMeters <= 100
- The DwC is intended to be simple as simplicity reduces the barriers for data providers and maximizes availability to data consumers, from desktop software to other analysis and integration services.
- The DwC is intended to accommodate data that are relatively common—concepts that can be used across more than one discipline, concepts that are commonly found in relevant databases, and populated frequently enough to be broadly useful.

At present the DwC V2 (Darwin Core Version 2) is only an unordered set of simple elements. Its simplicity allows straightforward configuration and easier interfaces (Canhos et al. 2004). However, its simplicity makes the DwC inadequate for some purposes (TDWG 2006).

- The DwC is not a sufficient model or data structure for managing primary data, such as a collection database.
- The DwC is also expected to be insufficiently complete for the needs of almost any specialized discipline of biodiversity. Data concepts deemed useful or

required by such community can be specified in an extension of the DwC—a new schema that imports the DwC and adds new elements for use within the specialized community.

While DwC does not satisfy the needs of full, detailed data exchange, DwC was conceived as a foundation for functional data exchange in the DiGIR and Species Analyst projects (Canhos et al. 2004). The DwC V2 data format is presented in Appendix B.

3.3.3.3 Taxonomic Concept Transfer Schema

The Taxon Concept Schema (TCS) was developed to enable the transfer of taxonomic name and concept data using an abstract model of taxonomic concepts which can capture the range of models represented and understood by the various taxonomic data providers. The model, presented as an XML schema document, has been through several revisions. TCS v1.0 was voted to be recommended as a TDWG standard (IUBS 2005).

3.4 Technical Aspects of Database Integration

The technical areas related to database integration were studied so that proper and suitable methods could be utilized to propose a new database integration system in this research. In the next subsections, technical factors that bind a database integration system are explained to the extent that they cover the scope of this research.

3.4.1 Integration Approaches

Integration approaches are important in the sense that they provide a scrupulous platform in which to work at building database integration systems. Four broad approaches for integrating databases are described in detail in the following subsections (see Table 3.3). These four approaches are illustrated in Fig. 3.5.

3.4.2 Web Pages and Interfaces

With the emergence of the Information and Communication Technology (ICT), most public databases are put online so that they can be accessed and searched via Web pages. In a database integration system, a Web portal is needed for users to

Table 3.3 Integration approaches

	Navigational	Warehouse	Mediator or federation	Indexing flat files
Definition	The idea of navigational or link-based integration emerged from the fact that an increasing number of sources on the Web require of users that they manually browse through several Web pages and data sources in order to obtain the desired information (Davidson et al. 1995). The specific paths comprises workflows in which the output(s) of a source or tool is (are) redirected to the input of next source until the requested information is reached (Buttler et al. 2002). Consequently, queries are transformed into (possibly several) path expressions that could each answer the query with different levels of satisfaction (Mork et al. 2001)	Warehouse integration consists in materializing the data from multiple sources into a local warehouse [Hammer and Schneider (2003) call it a Unifying Database] and executing all queries on the data contained in the warehouse rather than in the actual sources (Thomas and Subbarao 2004). Warehousing emphasizes data translation, as opposed to query translation in mediator-based integration (Sujansky 2001). Basically, in this method, data loaded from multiple biological sources are mapped into a standard data format before it is physically stored locally	Database integration systems that use mediation or federation typically consist of three elements: wrappers, an integration layer and a query interface. The wrappers provide uniform access to the heterogeneous data sources. The integration layer decomposes user queries, sends them to relevant wrappers and eventually integrates the query results before the result is returned to the user via the query interface. In contrast to mediated databases, in federated databases wrappers mainly map the different interface between the data sources, Thus in database federations it is required that the data sources provide the main search and query functionality via different interfaces. In many cases the data sources (Web pages and flatfiles) do not provide suitable search or query methods. In mediated databases, the wrappers play a more active role and when necessary implement missing search or query methods for the data sources (Jacob 2004)	The main purpose of indexing systems is simple: the databases to be integrated are provided as flatfiles. The integration system indexes these flatfiles using a script, which has to be provided for each database. The indexing script is also responsible for discriminating datatypes and for generating links to other relevant databases (Jacob 2004)

(continued)

Table 3.3 (continued)

Navigational	Warehouse	Mediator or federation	Indexing flat files
		Mediator based integration concentrates on query translation. A mediator in the information integration context is a system that is responsible for reformulating at run-time a query given by a user on a single mediated schema into a query on the local schema of the underlying data sources. None of the data in the mediator-based integration system is converted to a unique format according to a data translation mapping. Instead a different mapping is required to capture the relationship between the source descriptions and the mediator to be translated to queries on the data sources. The author then added that, specifying this correspondence is a crucial step in creating a mediator, as it will influence both how difficult the query reformulation is and how easily new sources can be added to or removed from the integration system (Hernandez and Kambhampati 2004)	Based on the indexes, users can search several indexed databases in one step. When indexing systems discriminate between datatypes, the databases can also be searched using common datatype specific comparison operators. In addition, indexing systems can be searched using sequence similarity search tools and the results are visualized in an appropriate way

(continued)

Table 3.3 (continued)

	Navigational	Warehouse	Mediator or federation	Indexing flat files
Advantages	Eliminates relational modeling of the data and instead applies a model where sources are defined as sets of pages with their interconnections and specific entry-points, as well as additional information such as content, path constraints, and optional or mandatory input parameters (Cali et al. 2002; Lenzerini 2002) Effectively allows the representation of cases where the page containing the desired information is only reachable through a particular navigation path across other pages	Does not rely network for data access, problems such as network bottlenecks, low response times and occasional unavailability of sources can be eliminated. This is because query sis performed locally. (Husain et al. 2004) Using materialized warehouses allows for an improved efficiency of query optimization as it can be performed locally (Florescu et al. 1998; Davidson et al. 1995) Allows the system/user to filter, validate, modify and annotate the data obtained from the sources (Davidson et al. 2001, Hammer and Schneider 2003), and this has been noted as a very attractive property for bioinformatics (Thomas and Subbarao 2004) Restricted to integrating only a few source databases, but manage to achieve a higher degree of interoperability on the integrated data sources (Jacob 2004)	User does not need to know where the data is stored Source databases transparently deliver up to date results	Support various flatfile formats addition or removal of any number of flatfile databases without affecting other databases in the integration system (Jacob 2004)

(continued)

Table 3.3 (continued)

	Navigational	Warehouse	Mediator or federation	Indexing flat files
Disadvantages	Does not support bulk queries	Possibility of returning outdated results (Thomas and Subbarao 2004) Must regularly check throughout the underlying sources for new or updated data and then reflect that modification on the local copy of the data (Davidson et al. 1995) and this can be difficult and sometimes impossible Limited due to the fact that it is generally not possible to integrate new databases without changing the schema of data warehouse, because this is usually associated with conflicts between the schemas of the new database and the data warehouse schema (Jacob 2004) Creates redundancy as there are records in the distributed database as well as in the local database therefore can be time consuming	Not efficient if data is parsed through html pages and/or is the user poses a query which due to restricted interfaces, cannot be directly passed to the source databases Speed to access data relies on the network traffic. Since the query is executed at the data source, the query response time depends on the usage access of data sources (Husain et al. 2004)	Does not provide a mechanism to integrate in-house relational databases, nor does it provide a mechanism to perform data cleaning and transformation for complex data mining (see http://www.iscb.org/ismb2000/tutorials/griffiths.html)
Use	Best for integration of sources which provide the users with pages that would not–or hardly- be accessible without point-and-click navigation	Best for integration of limited data sources with stable database schemas	Best for integration of large amount of data locally or when advanced methods for data processing and analysis have to be integrated (Jacob 2004)	De facto standard for the integration of high numbers of heterogeneous databases

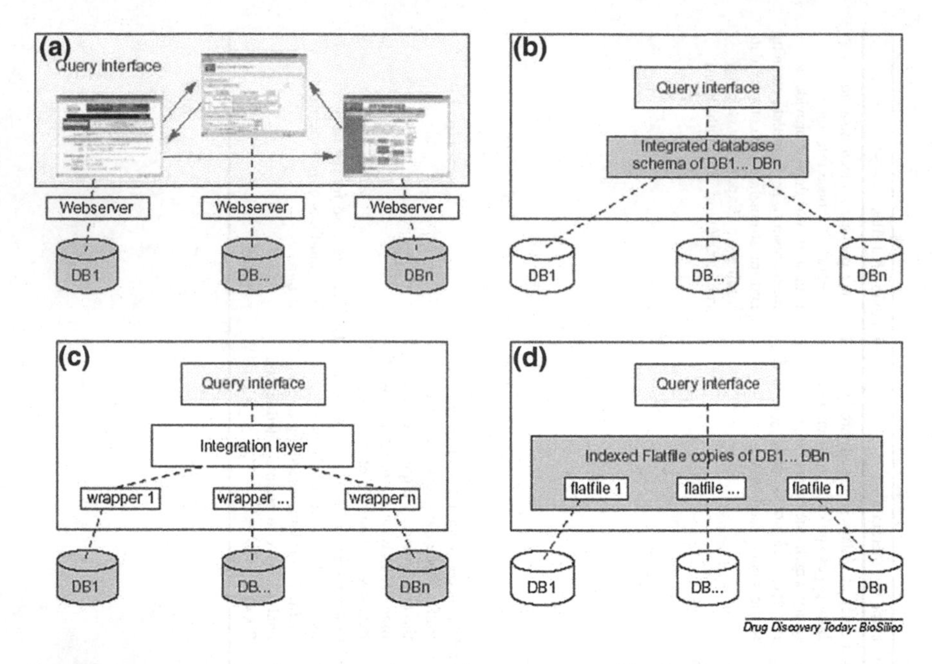

Fig. 3.5 Overview of database integration architectures, exemplified by the integration of n data sources (*DB*1 to *DBn*). The localization of the data against which user queries are processed is depicted in *orange*. *DB*1, DB and the database to be integrated. **a** Hypertext navigation. **b** Data warehouse. **c** Database mediation and federation. **d** Indexing flat files. (Adapted from Jacob 2004)

search and retrieve data from the integrated remote databases. Therefore, Web pages and interfaces are discussed in this section.

The development of Hypertext Markup Language (HTML) has spurred a tremendous growth of the Internet over the past 5 years. The original HTML language, allowed only hyperlinks and some basic formatting of ASCII text. Later HTML is used as a language to create forms which allows the client to transmit data to the browser (Hatzigeorgiu and Syropoulos 1999).

Hatzigeorgiu and Syropoulos (1999) defined HTML as an interpreted markup language based on tags within regular ASCII text. The Web browser contains an interpreter for this language and a viewer for resulting formatted text. The HTML documents (Web pages) are being kept on a Web server. The Web server and browser communicate with each other using the HTTP protocol. The client–server model of the HTTP protocol is simple: the client requests a document; the server sends the document to the client and closes the connection.

Hatzigeorgiu and Syropoulos (1999) also mentions that the simplicity of the original form of both the HTML language and the client–server model of the HTTP protocol are largely responsible for the success of the Web. With this, writing a Web page has become an uncomplicated task. Hatzigeorgiu and Syropoulos (1999) also highlighted that the simplicity is the major problem in developing Web applications. Some of the shortcomings are:

- HTML is a markup language, not a programming language. It does not even offer some type of loop construct.
- HTTP requests from the client can only be for "whole" documents, not for "chunks" of data. Thus, the client can only show or change a whole document, not parts of it.
- HTTP is stateless. After the server has fulfilled a client request, no memory of the transaction is kept.

Therefore, it is quite difficult to develop dynamic Web applications under these restrictions. Dynamic content is contrasted to the static content of the regular Web pages. It can be the result of an interaction with a database of some short, which leads to the construction of an HTML page. Usually this is achieved by server-side programming techniques.

Besides that, in a static Web page, content is determined when the page is created. As users access a static page, the page always displays the same information. In a dynamic Web page, content varies based on user input and data retrieved from external sources. It is vital to distinguish between data-based Web pages from pages created through client-side scripting technologies, such as JavaScript and VBScript, that support such tasks as verifying data, displaying new browser windows, and providing animated graphics and sound, rather than interacting with files or databases (Morrison et al. 2002).

As reported by Morrison et al. (2002), a data-based Web page is requested when a user clicks a hyperlink or submit button on a Web page form. If the request comes from clicking a hyperlink, the link specifies either a Web server program or a Web page that talks to a Web server program. Although this query requires no user input, the results vary depending on when the query is made. If the request is generated when the user clicks a Web page form's submit button, instead of a hyperlink, the Web server usually uses the form inputs to create a query. In either case (submit button or hyperlink), the Web server is responsible for formatting the query results by adding HTML tags. The Web server program then sends the program's output back to the client's browser as a Web page.

In a Web page, client-side scripts can be embedded into HTML to create more sophisticated interfaces. Client side scripts such as JavaScript and VBScript are embedded in a HTML document, along with the static HTML text. It is placed within delimiter tags to indicate to the user's browser that the text is code rather than Web page text. If the user's browser is able to recognize and interpret the code, it is processed. If the browser is unable to recognize and interpret the code, it is displayed as text on the Web page, unless the script author encloses it in a HTML comment.

However, as stated by Morrison et al. (2002), client side scripts cannot be used by a Web page to interact with remote databases, they are often used to validate user inputs entered on HTML forms submitted for processing by a server-side program. For example, a script running on a client workstation might check the inputs users submit to a Web page to make sure they entered all required data and appropriate data values. This approach avoids transmitting inputs to the Web server that are incomplete or include errors, while offloading error checking and handling from the Web server program to the client workstation.

Morrison et al. (2002) also argue that client side scripts can also be used to create advanced Web page features such as:

- Image maps allowing users to move their cursors over an image and click to access different Web page links.
- A new browser window displaying an alternate Web page.
- A timer displaying an image or playing a sound.
- Cookies storing data on users' computers about their actions while browsing a particular Web page.

JavaScript is the most commonly used client-side scripting language and it is supported by most browsers. It can also support server-side processing through Web-server specific languages derivatives; a version of JavaScript called LiveWire can be used on Netscape Web servers for server-side processing of form inputs; and Jscript, a Microsoft—specific version of JavaScript, can be used in Active Server Pages (ASP) pages on Microsoft Web Servers. Although JavaScript's core language is similar to Java's core language, JavaScript uses loosely typed variables and is designed to work mainly within Web browsers. JavaScript client-side scripts can be added to standard HTML Web pages through special HTML tags.

There are many aspects that are linked to client side scripting languages such as user operating system, browser platforms, and developer expertise. If a particular Web page is to be accessed by a variety of users over the Internet, JavaScript is probably better than VBScript, because JavaScript is the only scripting language able to run on nearly all browsers. If the Web pages are to be accessed on an intranet and if the organization uses, as its standards, Microsoft's browser and Web server, VBScript is a satisfactory scripting language for creating client-side scripts.

In view of the fact that the client side scripts cannot be used by a Web page to interact with remote databases, there has to be a method for the database to communicate with the Web. This is further discussed in the next section.

3.4.3 Database Integration with the Web

Communication between heterogeneous databases requires a database to be accessible via the Web. Web server is required to accomplish this. The Internet Information Server (IIS) comes with Windows XP Professional or Windows 2003 Server package whereas Apache comes from The Apache Software Foundation (2006), or commercial products (Muse 2006). The core function of the Web server is to handle the Hypertext Transfer Protocol (HTTP) request and response. It is required for providing production services (or example, a database) to clients through the World Wide Web.

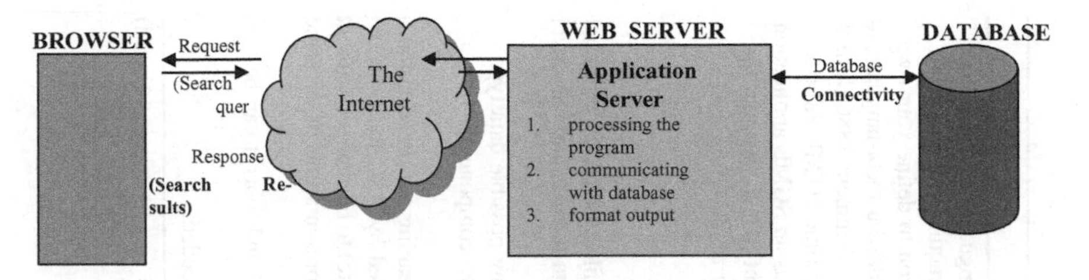

Fig. 3.6 A process for a client to get information from a database through the web (adapted from Xia et al. 2002)

The process of searching a database can be summarized in the points below (Xia et al. 2002):-

- User logs on to the Web site by typing the Uniform Resource Locator (URL) in the browser for example the Microsoft's Internet Explorer.
- The user types the request in the search and submits the request.
- The server program then communicates to the database using Structured Query Language (SQL) through the Open Database Connectivity (ODBC) and Java Database Connectivity (JDBC), obtains and formats the output, and sends it back to the user (see Fig. 3.6).

However, the above process is applicable and sufficient only for searching a single database at any one time. To search multiple heterogeneous databases (which is the focus of this research), a mediator has to be embedded in the system (see Fig. 3.6). The implementation of mediator will be discussed in detail in this research.

3.4.4 Markup Languages for Database Integration System

Choosing the right programming language is crucial for the success of any software or application development project. One technology may require a longer learning curve than another. Some may be more suitable for handling complicated tasks than others (Morrison et al. 2002).

The database integration system discussed in this research is Web based. Therefore, it is important to study the markup languages needed to build the database integration system.

XML and HTML are known to be essential markup languages for database integration. A brief explanation about HTML was given in Sect. 3.4.2. In this section, HTML is discussed in relation to XML (see Table 3.4).

XML is known to be a special language for database integration. It can be used to specify semantics associated with the contents of databases. This is further discussed in Sect. 3.4.4.1.

Table 3.4 Features of XML and HTML

	HTML	XML
Definition	HTML is an interpreted markup language based on tags within regular ASCII text. The Web browser contains an interpreter for this language and a viewer for the resulting formatted text. The HTML documents (web pages) are being kept on a web server. The Web server and browser communicate with each other using the HTTP protocol. The client–server model of the HTTP protocol is simple: the client requests a document; the server sends the document to the client and closes the connection (Nick and Apostolos 1999) (see Sect. 2.4.2)	XML is derived from the Standard Generalized Markup Language (SGML), the international standard for defining descriptions of the structure and content of different types of electronic documents. SGML allows you to define your own markup language that consists of your own tags. It is indeed a meta-language to design markup languages. For example, the Request For Comment 1886 (http://www.ietf.org/rfc1866.txt) described formally for the first time HTML 2.0 (to distinguish it from the previous informal specifications) as an SGML application. XML could be described as a Web dedicated, lightweight SGML, without the more complex and less used parts of SGML (Frederic et al. 2001)
Advantages	User-friendly: browsing the Web is almost instinctive and requires very minimal training time Easy to learn: the knowledge of a few self-explanatory tag names and the understanding of a simple syntax is enough to write surprisingly good looking HTML pages Easy to program: the CGI mechanisms allow fast linking of programs to an HTML page	Embed existing data: mapping existing data structures like file systems or relational databases to XML is simple. XML supports multiple data formats and can cover all existing data structures Embed multiple data types: XML documents can contain any possible data type - from multimedia data (image, sound, and video) to active components (Java applets, ActiveX). Separates content from presentation: XML tags describe meaning not presentation. The look and feel of an XML document can be controlled by XSL style sheets, allowing the look of a document (or of a complete Web site) to be changed without touching the content of the document. Multiple views or presentations of the same content are easily rendered. Easy: information coded in XML is easy to read and understand, plus it can be processed easily by computers. No fixed set of tags: New tags can be created as they are needed.

(continued)

Table 3.4 (continued)

	HTML	XML
Disadvantages	Does not offer some type of loop construct	XML documents occupy more memory or disk space than binary representations due to the use of text for everything and the presence of the tags.
	Does not define the semantics of a document: for example, the <blockquote> tag announces a list, but no difference can be made in HTML between a list of restaurants in Kuala Lumpur and biodiversity of flora. The HTML META tags only indicate vaguely the content of an HTML page, are not widely used and lack standardization (Lawrence and Giles 1999). Therefore, it is not easy to programmatically extract information from HTML pages. Search engines produce number of incorrect hits for some types of queries because HTML is not capable of capturing the semantics of a document	It takes more work to read, write or transmit XML documents than binary formats. The tags must be read and processed, and information such as numbers will have to be converted from its textual form to the binary form that the application needs. When using XML as a database, accessing the data is slow due to parsing and text conversion, compare to conventional databases. Reading an XML document is more complicated than reading a fixed-format binary document due to the tag processing that must occur. Thus, it will be more difficult to create an XML parser than a parser for a particular binary format. However, an XML parser can be reused for many different applications, and they are freely available. XML documents must either be converted into HTML before distribution or converting it to HTML on-the-fly by middleware. Barring translation, developers must code their own processing applications
Use	Website development	Integration of Document Content (Stonebraker et al. 2001, Andrew and Paula 2005);
	Database browsing	
	Data publishing	
	Data gathering	Specifying business rules (Yang et al. 2003) in contracts (Grosof et al. 1999);
	Data submission	
	Data analysis	Generate Web-based data automatically (Turau 2002), (Lim and Wen 2002)
		Database interaction mark-up language (Han et al. 2000, Lim and Wen 2002);
		Vehicle for document sharing in virtual organisations (Yang et al. 2001);
		Integration of business catalogues into a single unified data structure (Stonebraker et al. 2001);
		Role-based access control of documents (Yang et al. 2001)

(continued)

Table 3.4 (continued)

	HTML	XML
Special properties	Forms which allow the client to transmit data to the browser tables Scripts allow the usage of interpreted high level language (such as JavaScript or Visual Basic Scripting Edition) on the client, provided that the client includes the necessary interpreter for the scripting language Frames allow simultaneous viewing of more than one hypertext document and consequently give the ability to change only a part of the viewable area.	Enforced and defined structure (XML rules and schemas) Formal metadata (through the ISO 11179 specification) Namespaces (permits sharing of uniquely identifiable metadata tags) Linking data via the internet Logic and meaning (the Semantic Web) Self-awareness (embedded protocols and commands)

3.4.4.1 XML as a Tool for Data Integration

Integrating heterogeneous data sources requires more than a tool for organizing data into a common syntax (Bertino and Ferrari 2001). The nature of data can be categorized as follows.

Data models: Data sources can greatly differ with respect to the structures they use to represent data (for example, tables, objects, files). Reconciliation of heterogeneous data models requires a common data model to map information coming from the various data sources.

Data schema: Once one has agreed upon a common data model, the problem arises of reconciling different representations of the same entity or property. For example, two sources may use different names to represent the same concept ("species" and "spesies"), or the same name to represent different concepts or two different ways for conveying the same information ("date of birth" and "age"). Additionally, data sources may represent the same information using different data structures. The need thus arises for tools to reconcile all these differences.

Data instances: At the instance level, integration problems include determining if different "objects" coming from different sources represent the same real-world entity and selecting a source when contradictory information is found in different data sources (for instance, different birth dates for the same person).

Given these issues concerning data integration, it is important to understand whether and how XML can help the integration process and where XML is not enough. Bertino and Ferrari (2001) explain this in the following points.

- *Data Model Reconciliation*
  XML provides a relatively natural way of structuring data, based on hierarchical, graph-based representations. Such models have the advantage of being simple, standard, and well accepted, but also powerful enough to represent structured, unstructured, and semi structured information. Thus, XML works well as a common data model over the Web, and a variety of tools have been marketed to support database contents publishing in XML. In addition, several research efforts have investigated the use of graph-based data models for managing semi structured data and for integrating heterogeneous data sources.
- *Schema Reconciliation*
  The names and meanings of XML tags are arbitrary, which makes it critical to agree on a set of standardized domain specific tags and schemas. Standardization of tags and schemas would greatly simplify the task of data integration, but a way to describe the semantics of elements and attributes is also needed.
- *Instance Reconciliation*
  Metadata play a crucial role in instance reconciliation, determining how to deal with similar or contradictory information.

A combination of both HTML and XML can be used to build a dynamic database integration system. The use of these languages is explained in this chapter of this book.

3.4.5 Scripting Languages

A database integration system requires the development of a dynamic Web portal that will send queries to the backend database and so forth. In developing Web portals, besides markup languages, scripting languages are used as quick and secure method of data transfer across backend databases while providing security against unauthorized access from external internet users. Therefore, in building a Web portal connected to mediators (which connect integrated heterogeneous backend databases), there is a need to use a suitable scripting language.

In this section, four scripting languages are discussed. They are Active Server Pages (ASP), PHP Hypertext Preprocessor (PHP), Java Server Pages (JSP) and ColdFusion.

ASP and PHP are different approaches to building dynamic Web sites that can incorporate database interactivity and other application server uses into Websites.

ASP is used with “Internet Information Server” (IIS) which is a program that runs on Microsoft servers. ASP is generally not supported beyond Microsoft servers yet widely used for large companies Web need (Brown 2006).

PHP is a widely-used Open Source general-purpose scripting language that is especially suited for Web development and can be embedded into HTML. Its syntax draws upon C, Java, and Perl, and is easy to learn. The main goal of the language is to allow Web developers to write dynamically generated Web pages quickly, but one can do much more with PHP (Achour et al. 2006).

PHP succeeds an older product, named PHP/FI. PHP/FI was created by Rasmus Lerdorf in 1995, initially as a simple set of Perl scripts for tracking accesses to his online resume. He named this set of scripts ‘Personal Home Page Tools’. As more functionality was required, Rasmus wrote a much larger C implementation, which was able to communicate with databases, and enabled users to develop simple dynamic Web applications. Rasmus chose to release the source code for PHP/FI for everybody to see, so that anybody can use it, as well as fix bugs in it and improve the code (Achour et al. 2006).

JSP (Sun Microsystems 2006) is an extension to the Java servlet technology pioneered by Sun. Like ASP and PHP it provides a simple programming vehicle for displaying dynamic Web content. JSP is the Sun/Java counterpart to Microsoft’s ASP. JSP is not widely used by Web developers.

The first version of ColdFusion (then called Cold Fusion) was released in 1995, written almost entirely by one man, JJ Allaire. All versions of ColdFusion prior to 6.0 were written using Microsoft Visual C++. This meant that ColdFusion was largely limited to running on Microsoft Windows, although Allaire did manage to port ColdFusion to Sun Solaris starting with version 3.1.0. ColdFusion is an application server and software development framework used for the development of computer software in general, and dynamic Web sites in particular. The primary feature of ColdFusion is its associated scripting language, ColdFusion Markup Language (CFML), which compares to JSP or PHP and resembles HTML in syntax (Wikimedia Foundation 2006).

Table 3.5 Characteristics of scripting languages (adapted from Vasudevan 2001)

	Platform	Speed (how many pages per second it can pump out)	Flexibility	Web server
ASP	Windows	43 pages/second	Does not support beyond Microsoft servers	Internet information server
PHP	NT and Unix	47 pages/second	Easy to use and feature packed program that allows creation dynamic and database driven web sites effortlessly Supports open source	Apache
JSP	All platforms	13 pages/second	Simple programming vehicle for displaying dynamic web Not restricted to any platforms	Internet information server, apache
ColdFusion	Windows and Unix	29 pages/second	Incorporate legacy databases using COM objects Robust caching facilities Not restricted to any platform	Internet information server, apache

The Zdnet (Vasudevan 2001) did an evaluation and benchmarking of four Web scripting languages. During benchmarking, the same specifications and identical CPU, memory boxes were used. Under identical conditions, it was found that PHP was the fastest—about 3.7 times faster than JSP and about 1.2 times faster than ASP (see Table 3.5 for speed details).

The selection of a scripting language basically depends on the requirements of the software development. Variables such as operating systems and hardware requirements may influence the use of a particular language however, they are not as important as user's preference. They will evaluate the site's usefulness based on design, content, reliability and availability. Therefore as long as a website follows these rules, any of these dynamic scripting tools will work fine.

3.4.6 *Query Languages and Database Connectivity*

Query languages and database connectivity are two elements required to build a database integration system. Query languages are used to write queries which are sent to a Web server to communicate with a database. On the other hand, database connectivity is needed between a database and Web server to allow the Web server to process the query (see Fig. 3.7).

The typical paradigm for querying databases is through an expressive, declarative query language (such as SQL) that relies on database structure (Jennifer 1999). Therefore, a general approach for linking Web pages with a database involves creating a database connection, then issuing Structured Query Language (SQL) commands. The SQL commands cause the SQL queries to be executed and the records retrieved can then be displayed within a Web page. The code for establishing a database connection and issuing SQL command is language–and-database-specific (Morrison et al. 2002).

A study by Wong (2002) showed that the high-level query languages of the more general data integration systems surveyed are all SQL-like and thus designed to express traditional (nested relational) database-style queries.

As Database Management Systems (DBMS) now provide built-in access interfaces such as Java Database Connectivity (JDBC) and Open Database Connectivity (ODBC), it is possible to query the underlying databases using standardized query languages such as Structured Query Language (SQL) and Object Query Language (OQL), or grant access to life-science-specific search methods such as Basic Local Alignment Search Tool (BLAST) (Jacob 2004).

Java Database Connectivity (JDBC) which is mostly used to connect relational databases and consists of a set of classes and interfaces written in the Java, that can be used to address metadata, as well as data itself (Eaglestone and Ridley 2001), (Hamilton et al. 1997). JDBC provides a standard Application Programming Interface (API) for database developers and makes it possible to write database application using pure Java (Elbibas and Ridley 2003).

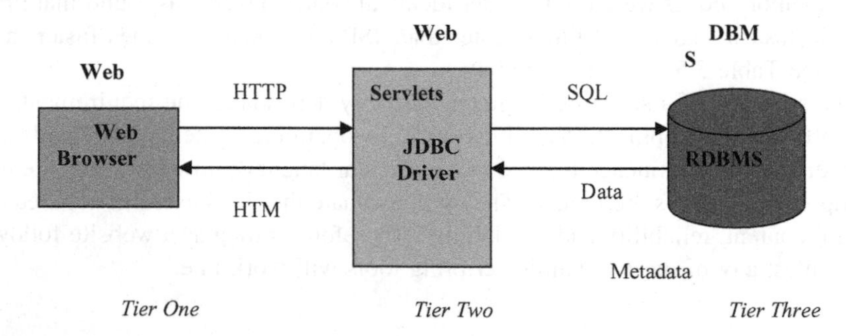

Fig. 3.7 Database accessibility solution via web. Three-tier solution (adapted from Elbibas and Ridely 2003)

3.4.6.1 Process of Database Connectivity via the Web

The process of database connectivity with the Web was briefly described in Sect. 3.4.6. In this section, the process is explained in detail.

Typically, database accessibility is divided into three layers as shown in Fig. 3.7 (Elbibas and Ridley 2003).

Tier one: the Web browser

The first tier uses a Web browser to take advantage of the installed user base of this universal client. An HTML form is used for user-input and the results of the database query are returned as an HTML page. These pages are not necessarily static.html files and may well be generated by the servlets (Hunter and Crawford 2001). Using HTML for user-input and displaying the data lowers the requirement of the client's browser version. This reduces the level of complexity required on the client's browser and has numerous advantages. It reduces the complexity of the Web applications wide deployment and it also makes the home user more confident as they will not have to download the latest Java run time environment or the latest browser version to run the Web application (Rajagopalan et al. 2002).

Tier two: the Servlet/Web Server

The second tier is implemented with a Web server running Java servlets. The Java servlet is able to access the database and return an HTML page. Servlets are runnable on most commercial Web servers (Hunter and Crawford 2001). An example of Web server is the (JSWDK) 1.0.1 Web server (Sun Microsystems 1999).

Tier three: the Database Server

The third tier is the back-end database server. The Java servlet can access information in the database provided that a JDBC driver exists. Java Database Connection (JDBC) implementations are available for the wide majority of databases.

Referring to the three tiers above, the basic steps of interaction in Fig. 3.7 can be described as follows:

1. The user enters information into an HTML form and it is passed to the Java servlet running on the Web server.
2. The Java servlet parses the form data and constructs an SQL statement. The SQL statement is passed to the database server using the Java Database Connection (JDBC).
3. The database server executes the SQL statement and returns a result set to the Java servlet.
4. The Java servlet processes the result set and constructs an HTML page with the data. The HTML page is then returned to the user's Web browser.

Database connectivity and query languages also depends on other attributes such as type of Database Management Systems (see Sect. 3.3.2) and the programming languages (see Sect. 3.4.5) used for the development of a system.

3.4.7 Web Services

The fundamental idea behind Web services is to integrate software applications as services. This concept is based on a defined set of technologies, supported by open industry standards, that work together to facilitate interoperability among heterogeneous systems, whether within an organization or across the Internet (Pullen et al. 2005).

The World Wide Web consortium (W3C) defines a Web service as follows:

A Web service is a software application identified by a URI, whose interfaces and binding are capable of being defined, described and discovered by XML artifacts and supports direct interactions with other software applications using XML based messages via Internet-based protocols (Austin et al. 2002).

Microsoft has two definitions for Web services:

1. A Web service is a unit of application logic providing data and services to other applications. Applications access Web services via ubiquitous Web protocols and data formats such as HTTP, XML, and SOAP, with no worries about how each Web Service is implemented (MSDN Library 2002).
2. A Web service is programmable application logic accessible using standard Internet protocols. Web service combines the best aspects of component-based development of the Web. Like components, Web services represent black-box functionality that can be reused without worrying about how the service is implemented. Unlike current component technologies, Web services are not accessed via object-model- (DCOM), Remote Method Invocation (RMI), or Internet InterORB Protocol (IIOP). Instead, Web services are accessed via ubiquitous Web protocols and data formats, such as hypertext Transfer Protocol (HTTP) and Extensible Markup Language (XML). Furthermore, a Web service interface is defined strictly in terms of the messages the Web service accepts and generates. Consumers of the Web service can be implemented on any platform in any programming language, as long as they can create and consume the messages defined for the Web service interface (Schnase 2000).

According to the above definitions, one point to be noted is that Web services do not necessarily exist on the Web. They can exist anywhere on the network, Internet and Intranet. Some Web services can be invoked by a simple method invocation in the same operating system process, or perhaps using shared memory between tightly coupled processes running on the same machine. In fact, Web services have little relationship with the browser-centric, HTML-focused Web (Graham et al. 2001).

Fundamentally, Web services provide an approach to distributed computing with application resources provided over networks using standard technologies. Because they are based on standard interfaces, Web services can communicate even if they are running on different operating systems and are written in different languages. The standards are widely supported by industry and have been applied successfully in a wide range of different domains. For this reason, Web services provide an excellent approach for building distributed applications that must incorporate diverse systems over a network (Pullen et al. 2005).

The Web services framework provides a set of operations, modular and independent applications that can be published, discovered, and invoked by using industry standard protocols such as the Extensible Markup Language (XML), Simple Object Access Protocol (SOAP), Web Service Description Language (WSDL), and Universal Description, Discovery and Integration (UDDI) (Pullen et al. 2005).

The main technology currently used for SOAP transport is the Hypertext Transfer Protocol (HTTP), but SOAP specification does not require HTTP. Although it might be able to use Simple Main Transfer Protocol (SMTP), File Transfer Protocol (FTP) or other methods, HTTP is currently the most common. SOAP is a lightweight and simple XML-based protocol that is designed to exchange structured and typed information on the Web (Box et al. 2000). The purpose of SOAP is to enable rich and automated Web services based on a shared and open Web infrastructure. WSDL offers an XML protocol for describing network services based on XML. UDDI project is an industry initiative that is working to enable businesses to quickly, easily and dynamically find and transact with one another. Essentially SOAP is the major processing protocol in Web Services. In other words, Web Services utilizes SOAP to process the service. The efficiency of SOAP will directly affect the efficiency of Web Services (Yu and Chen 2003).

3.5 Conclusion

In this chapter, five existing database integration systems were analyzed. Suitable features were identified from these five models and the generic characteristics were adopted to build the proposed database integration system in this research (see Sect. 3.2.6). Out of these five systems, DiGIR was selected as a model to integrate local biodiversity databases in the preliminary study. Because (1) DiGIR uses the Darwin Core data format which is the suggested data format to build the proposed database integration system (CABIR) (2) DiGIR is widely used especially by GBIF (Global Biodiversity Information Facility), and (3) DiGIR's installation files are easily available on the Internet. Other findings in this chapter include components and technologies to build biodiversity database integration system such as databases, DBMSs and data format and technical aspects related to database integration such as integration approach, programming languages, database connections, query language and Web service. These are further described in Chapter 4.

References

3rd Millennium Inc (2002). *Practical data integration in biopharmaceutical R&D: Strategies and technologies* (online), 125 Cambridge Park Drive, Cambridge, MA 02140. Available from: www.3rdmill.com/resources/3rd_Millennium_Biopharma-data-integration.pdf. Accessed 4 November 2005.

Achour, M., Betz, F., Dovgal, A., Lopes, N., Olson, P., Richter, G., Seguy, S., & Vrana, J., et al. (2006). *PHP manual* (online). Available from: http://www.php.net/manual/en/. Accessed 1 August 2006.

Andrew, B., & Paula, T. (2005). An XML-based architecture for data integration in vulnerability assessments. *Information Management and Computer Security, 13*(4), 260–273. Emerald Group Publishing Limited 0968-5227.

Anthony and Leslie, (2002). *The quest for data integration* (online). Incyte Genomics, Inc. Available from: http://www.incyte.com/insidegenomics/. Accessed 5 September 2005.

Austin, D., Barbir, A., & Garg, S. (2002). *Web services architecture requirements, W3C* (online). Available from: http://www.w3.org/TR/2002/WD-wsa-reqs-20020429. Accessed 10 January 2006.

Berendsohn, W. (2006). *ABCD schema* (online). Taxonomic department of biodiversity informatics and laboratories of the botanic garden and botanical museum Berlin-Dahlem. Available from: http://bgbm3.bgbm.fu-berlin.de/TDWG/CODATA/Schema/default.htm. Accessed 18 January 2005.

Bertino, E., & Ferrari, E. (2001). XML and data integration. *IEEE Internet Computing*, 089-7801/01 November December 2001.

Biodiversity Research Center (2005). *Distributed generic information retrieval* (online). Natural history museum. Available from: http://www.specifysoftware.org/Informatics/informaticsdigir/. Accessed 10 January 2005.

Box, D., Ehnebuske, D., Kakivaya, G., Layman, A., Mendelsohn, N., Nielsen, H. F., Thatte, S., & Winer, D. (2000). *Simple object access protocol* (*SOAP*) 1.1 (online). MIT computer science and artificial intelligence laboratory (CSAIL) U.S.A. Available from: http://www.w3.org/TR/2000/NOTE-SOAP-20000508/. Accessed 2 February 2006.

Brown, C. E. (2006). *ASP versus PHP: Which one is right for you* (online). Available from: http://www.pointafter.com/Archives/n10203.htm. Accessed 10 April 2005.

Buttler, D., Coleman, M., Critchlow, T., Fileto, R., Han, W., Liu, L., et al. (2002). Querying multiple bioinformatics data sources: Can semantic web research help? *ACM SIGMOD Record, 31*(4), 59–64.

Cali, A., Calvanese, D., de Giacomo, G., & Lenzerini, M. (2002). On the expressive power of data integration systems. *Proceedings of the 21st International Conference on Conceptual Modeling (ER 2002).*

Canhos, V. P., Souza, S., de Giovanni, R., & Canhos, D. A. L. (2004). Global biodiversity informatics: Setting the scene for a "New World" of ecological modeling. *Biodiversity Informatics, 1*, 1–13.

Davidson, S., Overton, C., & Buneman, P. (1995). Challenges in integrating biological data sources. *Journal of Computational Biology, 2*, 557–572.

Davidson, S., Crabtree, J., Brunk, B., Schug, J., Tannen, V., Overton, C., et al. (2001). K2/Kleisli and GUS: Experiments in integrated access to genomic data sources. *IBM Systems Journal, 40*(2), 512–531.

Eaglestone, B., & Ridley, M. (2001). *Web database systems*. London, UK: McGraw-Hill. ISBN 0077096002.

Elbibas, A. A., & Ridely, M. J. (2003). Using metadata for developing automated web system interface. *Copyright held by author.*

Elmasri, R., & Shamkant, B. (1994). *Fundamentals of Database* (2nd ed., p. 873). Reading, MA: Addison-Wesley. ISBN 0805317481.

Federal Geographic Data Committee (2000). *Biological data working group.* U.S. Geological Survey, 590 National Center Reston, Virginia, 20192.

Florescu, D., Levy, A. Y., & Mendelzon, A. O. (1998). Database techniques for the world-wide web: A survey. *ACM SIGMOD Record, 27*(3), 59–74.

Frederic A, Guy V, Emmanuel B (2001) XML, bioinformatics and data integration (Vol. 17(2), pp. 115–125). Oxford: Oxford University Press.

GBIF (2004). *Global biodiversity information facility* (online), Copenhagen, Denmark. Available from: www.gbif.org. Accessed 20 November 2005.

Graham, S., Simeomov, S., Boubez, T., Daniels, G., Davis, D., Nakamura, Y., et al. (2001). *Building web services with Java.* Indiana: SAMS.

Grosof, B.N., Labrou, Y., & Chan, H.Y. (1999). A declarative approach to business rules in contracts: Courteous logic programs in XML. *Proceedings of the First ACM Conference on Electronic Commerce*. ACM Press, New York.

Haas, L. M., Schwarz, P. M., Kodali, P., Kotlar, E., Rice, J. E., & Swope, W. C. (2001). Discovery link: A system for integrated access to life sciences data sources. *IBM Systems Journal, 40*, 489–511.

Hall, M. (2002). *Vendor profile: IBM in the Life Sciences*. Bio-IT-Infrastructure. 27549.

Hamilton, G., Cattell, R., & Fisher, M. (1997). *JDBC database access with Java: A tutorial and annotated reference*. Reading, MA: Addison Wesley. ISBN 0201309955.

Hammer, J., & Schneider, M. (2003). Genomics algebra: A new, integrating data model, language and tool processing and querying genomic information. *Proceedings of the 2003 CIDR Conference*.

Han, R., Perret, V., & Naghshineh, M. (2000). WebSplitter: A unified XML framework for multi-device collaborative web browsing. *Proceedings of the 2000 ACM Conference on Computer Supported Cooperative Work*. ACM Press, New York.

Hatzigeorgiu, N., & Syropoulos, A. (1999). New technologies for rapid development of web orientated database applications. *ACM SIGCUE Outlook, 27*(1), 25–31.

Hernandez, T., & Kambhampati, S. (2004). Integration of biological sources: Current systems and challenges ahead. *ACM SIGMOD Record, 33*(3), 51–60.

HGMIS (2005). Human genome management information system (online). U.S. Department of Energy Human Genome Program. Available at: http://www.ornl.gov/sci/techresources/Human_Genome/project/about.shtml. Accessed 15 December 2005.

Hunter, J., & Crawford, W. (2001). *Java Servlet Programming* (2nd ed.). USA: O'Reilly and Associates. ISBN 1 -56592-391-x.

Husain, W., Abdullah, R., Salam, R.A., & Zainol, Z. (2004). Development of a web-based system for integrated access to biological databases. *Transactions of Engineering, Computing and Technology*. ISSN 1305-5313.

IABIN (2004a). *Inter-American biodiversity information network* (online). Available from: http://www.iabin.net/english/index.shtml. Accessed 8 November 2005.

IABIN (2004b). *Biodiversity informatics*: *Information standards and metadata formats* (online). Available from: http://www.iabin.net/english/bioinformatics/protocols/standards_&_formats.shtml. Accessed 15 January 2006.

InforMax (2001) *GenoMax* (online), United States. Available from: http://www2.informaxinc.com/solutions/genomax/index.html. Accessed 15 December 2005.

International Union of Biological Sciences (IUBS) (2005). *Taxonomic databases working group: Time limited ballot* (online), Natural History Museum, London, United Kingdom. Available from: http://www.tdwg.org/TDWGStandardsBallot2005.htm. Accessed 20 November 2005.

International Working Group on Taxonomic Databases. (2005). *International union for biological sciences taxonomic database working group* (online), TDWG standards. Available from: http://www.tdwg.org/TDWGStandardsBallot2005.htm. Accessed 12 October 2005.

Jacob, K. (2004). Integration of life sciences databases. *DDT: BIOSILICO, 2*(2), 61–69.

Jennifer W (1999) Database management for XML. *IEEE Data Engineering Bulletin*, Special Issue on XML. *22*(3), 44–52.

Jones, A.C., Xu, X., Pittas, N., Gray, W.A., Fiddian, N.J., White, R.J., Robinson, J.S., Bisby, F.A., & Brandt, S.M. (2000). SPICE: A flexible architecture for integrating autonomous databases to comprise a distributed catalogue of life. To appear in *Proceedings of 11th International Conference and Workshop on Database and Expert Systems Applications* (*DEXA* 2000) (Lecture Notes in Computer Science). New York: Springer.

Lawrence, S., & Giles, C. L. (1999). Accessibility of information on the web. *Nature, 400*, 107–109.

Lenzerini, M. (2002). Data integration: A theoretical perspective. *ACM Symposium on Principles of Database Systems (PODS).*

MSDN Library (2002). *XML web services* (online). Available from: http://msdn.microsoft.com/library/default.asp?url=/nhp/default.asp?contentid=2800442. Accessed 10 July 2006.

Lim, B. B. L., & Wen, H. J. (2002). The impact of next generation XML. *Information Management and Computer Security Journal, 10*(1), 33–40.

Limi, A., Runyan, A., Anderson, V. (2005). *Darwin core 2* (online), Taxonomic Databases Working Group (TDWG). Available from: http://darwincore.calacademy.org/Documentation/PurposeGoals. Accessed 12 January 2005.

Mork, P., Halevy, A., & Tarczy-Hornoch, P. (2001). A model for data integration systems of biomedical data applied to online genetic databases. *Proceedings of the Symposium of the American Medical Informatics Association.*

Morrison, M., Morrison, J., & Keys, A. (2002). Integrating web sites and databases. *Communications of the ACM, 45*(9), 81–86.

Muse, D. (2006). *Server Watch* (online), Darien, Connecticut. Available from: http://www.serverwatch.com/stypes/index.php/d2Vi .Accessed 20 February 2006.

Nick, H., & Apostolos, S. (1999). New technologies for rapid development of Web oriented database applications. *ACM SIGCUE Outlook., 27*(1), 25–31.

Philippi, S. (2004). Light-weight integration of molecular biological databases. *Bioinformatics, 20*(1), 51–57. Oxford University Press.

Povey, S., Lovering, R., Bruford, E., Wright, M., Lush, M., & Wain, H. (2001). The HUGO Gene Nomenclature Committee (HGNC). *Hum Genet, 109*, 678–680.

Post, G. V. (2005). *Database management systems; designing and building business applications.* New York: McGraw Hill.

Pullen, J. M., Brunton, R., Brutzman, D., Drake, D., Hieb, M., & Morse, K. L. (2005). Using web services to integrate heterogeneous simulations in a grid environment. *Future Generation Computer Systems, 21*(1), 97–106.

Rajagopalan, S., Rajamani, R., Krishnaswamy, R., & Vijendran, S. (2002). *Java servlet programming Bible.* USA: Hungry Minds Inc. ISBN 0-7645-4839-5.

Schnase, J. (2000). Research directions in biodiversity informatics. *Proceedings of the 26th VLDB Conference.* Cairo, Egypt.

Siegel, J. (1998). OMG overview: CORBA and the OMA in enterprise computing. *Communications of the ACM, 41*(10), 37–43.

Species2000 (2000). *Indexing the world's known species* (online), University of Reading, UK. Available from: http://www.sp2000.org/about. Accessed 15 September 2005.

Stonebraker, M., Joseph, M., & Hellerstein, J.M. (2001). Content integration for e-business. *Proceedings of the 2001 ACM SIGMOD International Conference on Management of Data.* ACM Press, New York.

Sujansky, W. (2001). Heterogeneous database integration in biomedicine. Methodological review. *Journal of Biomedical Informatics, 34*(4), 285–298.

Sun Microsystems, Inc. (1999). *JavaServerTM web development Kit, Version 1.0.1* (online). Available from: http://java.sun.com/products/servlet/README.html. Accessed 20 November 2004.

Sun Microsystems, Inc. (2006). *JavaServer pages technology documentation* (online), U.S.A. Available from: http://java.sun.com/products/jsp/docs.html. Accessed 15 April 2006.

Taxonomic Databases Working Group (TDWG) (2006). *Darwin core 2* (online), Bournemouth University. Available from: http://darwincore.calacademy.org/Documentation/PurposeGoals. Accessed 15 August 2006.

The Apache Software Foundation (2006). *Apache* (online), U.S.A. Available from: http://www.apache.org.my .Accessed 15 April 2006.

Thomas, H., & Subbarao, K. (2004). Integration of biological sources: Current systems and challenges ahead. *ACM SIGMOD Record, 33*(3), 51–60.

Turau, V. (2002). Web and e-business application: a framework for automatic generation of Web-based data entry applications based on XML. *Proceedings of 2002 ACM Symposium on Applied Computing*. ACM Press, New York.

Vasudevan, A. (2001). *PHP how-to* (online). Available from: http://www.linuxdocs.org/HOWTOs/PHP-HOWTO-13.html. Accessed 19 December 2005.

Vieglais, D. (2003). *The Darwin core* (online) SpeciesAnalyst. Available from: http://speciesanalyst.net/docs/dwc/index.html. Accessed 12 January 2005.

Wikimedia Foundation Inc. (2006). *ColdFusion* (online), Florida. U.S.A. Available from: http://en.wikipedia.org/wiki/ColdFusion. Accessed 10 April 2006.

Wong, L. (2000a). Kleisli, a functional query system. *Journal of Functional Programming, 10*(1), 19–56.

Wong, L. (2000b). Kleisli, its exchange format, supporting tools, and an application in protein interaction extraction. *Proceedings of IEEE International Symposium Bio-Informatics and Biomedical Engineering*, pp. 21–28.

Wong, L. (2002). Technologies for integrating biological data. *Briefings in Bioinformatics, 3*, 389–404.

Xia, Y., Stinner, R.E., Chu, P.C. (2002). Database integration with the web for biologist to share data and information. *EJB Electronical Journal of Biotechnology* ISSN: 0717-3458.

Yang, J., Van den Heuvel, W.J., & Papazoglou, M.P. (2001). Service deployment for virtual enterprises. *Australian computer science communications. Proceedings of the Workshop on Information Technology for Virtual Enterprises*. ACM Press, New York.

Yang, C., Chang, N., & Zhang, C. N. (2003). An XML-based administration method on role-based access control in the enterprise environment. *Journal of Information Management and Computer Security, 11*(5), 249–257.

Yu, S. C., & Chen, R. S. (2003). Web services: XML-based system integrated techniques. *Emerald-The Electronic library, 21*(4), 358–366. MCB UP Limited.

Chapter 4
Methodology

4.1 Introduction

The main focus of this chapter is to build a new database integration system for retrieval of biodiversity information. This system which is named as **C**ommunication **A**rchitecture for **B**iodiversity **I**nformation **R**etrieval (CABIR) is based on the models presented in Chapter 3. In this chapter, the materials and methods to build CABIR are described in detail (see Fig. 4.1). These methods and materials are based on the findings in Chapter 3.

Building a database integration system requires relational biodiversity databases as ingredients to test the system. In view of this new relational databases were developed using primary biodiversity data sets gathered by local scientists. Prior to building these databases, a data format was proposed using methods described in Sect. 4.2.1. The attributes in the proposed data format were used to build relational biodiversity databases in Sect. 4.2.2.

The development of CABIR consists of three major steps which are database integration, building the wrappers and validation using XML and finally developing the search engine and interface (see Fig. 4.1). Elements such as DBMSs, integration approach, programming languages, database connections, query languages and Web services which were used to build CABIR are discussed in detail in this chapter (see Chapter 3).

4.2 Methodology for Biodiversity Data Format and Development of Relational Biodiversity Databases

Biodiversity data sources are relational databases containing primary biodiversity data. A standard data format is required to build these databases. Thus requirements analysis for a standard data format was done and the details are described

S. K. Dhillon and A. S. Sidhu, *Data Intensive Computing for Biodiversity*,
Data, Semantics and Cloud Computing 485, DOI: 10.1007/978-3-642-38047-1_4,

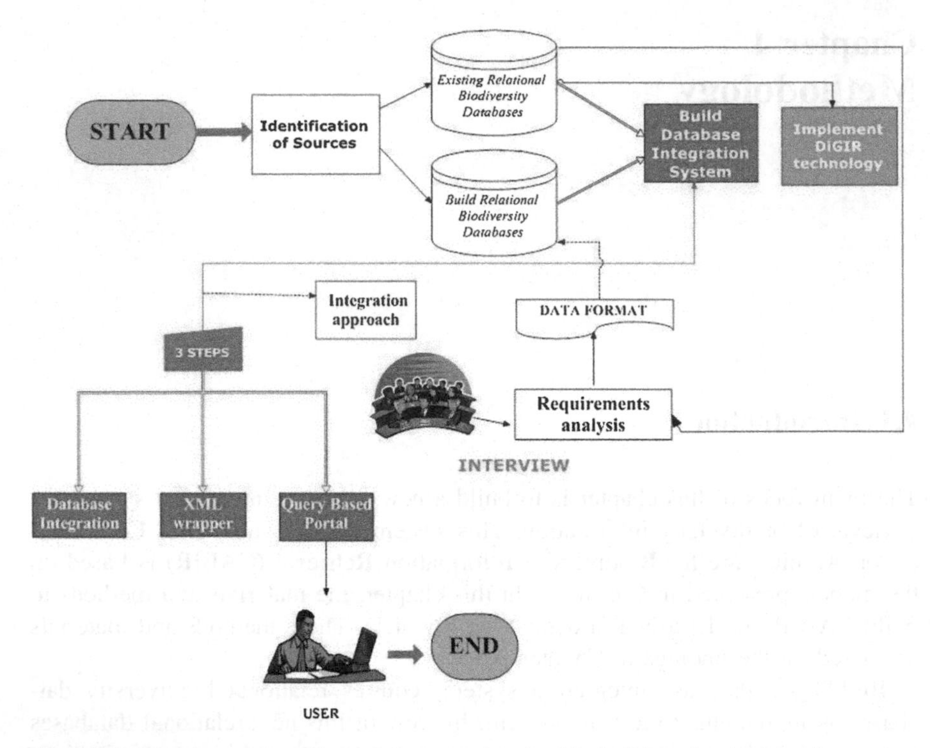

Fig. 4.1 Schematic representation of processes involved in building the database integration system (CABIR) in this research (*blue* represents CABIR and *green* represents DiGIR)

below. Hence after, the methodology for development of new relational biodiversity databases is described in detail.

4.2.1 Requirements Analysis for a Biodiversity Data Format

Integration of a biological database with the Web provides universal access to a database. However, to achieve universal data sharing, standardization is vital (Xia et al. 2002). Standardization involves a standard terminology for data storage and exchange. Sect. 2.3 in Chapter 2 described the global standards for sharing biodiversity data.

In this research, the Darwin core V2 was used as a model for gathering information on the data format for constructing databases in next section. Experts in the areas of parasites, fern, ginger, microorganisms and fish were interviewed. The results of the interview were used as attributes that are needed in a biodiversity database. The questionnaire used for the interview is attached in Appendix A. The data in the questionnaire was based on Darwin Core V2 data fields (see Sect. 3.3.3.2).

In addition to the questionnaire-based analysis, a study was also performed on existing biodiversity databases. The Malaysian Indigenous Microbial Online Databases (MIMODS, Merican 2005) consisting the Malaysian Indigenous Protozoa (MIP), Malaysian Indigenous Fungi (MIF), Malaysian Indigenous Virus (MIV), Malaysian Indigenous Algae (MIA) and Malaysian Indigenous Bacteria (MIB) were used to analyze their data formats. The results of both the analysis are presented in Chapter 4.

4.2.2 Development of Relational Biodiversity Databases

The process of building databases in this research is summarized in the following steps.

4.2.2.1 Digitization of Data and Images

Most biodiversity data is stored in journals, books and reports. Therefore, the first step was to transform the data into digitized form.

The primary sources of data for the relational databases were obtained from Prof. Susan Lim Lee Hong (parasites), Prof. Haji Muhammad (fern), Prof. Halijah Ibrahim (zingiberaceae- ginger), Assoc. Prof. Amir Feisal Merican (microorganisms) and Dr. Chua Tock Hing (fruit flies).

4.2.2.2 Naming of Database Fields

The fields for biodiversity databases were named according to the data format which is presented in Sect. 5.2 (see Sect. 4.2.1 for the method of analysis). As an example, the field names in the Host-Parasite relational database are presented in Fig. 4.2.

4.2.2.3 Designing Tables and Relationships

Next, the fields are assembled to form tables in a database. The tables and relationships among them are described using an Entity-Relationship Diagram (E-R) (see Fig. 4.3).

Data normalization is the process of creating a well-behaved set of tables to efficiently store data, minimize redundancy, and ensure data integrity. It was performed on the relational databases to minimize duplication of records (Post 2005). The tables are related to one another through primary and foreign keys. Primary key is a column or set of columns that identifies a particular row in a table whereas foreign key is a column in one table that is a primary key in a second table (Post 2005).

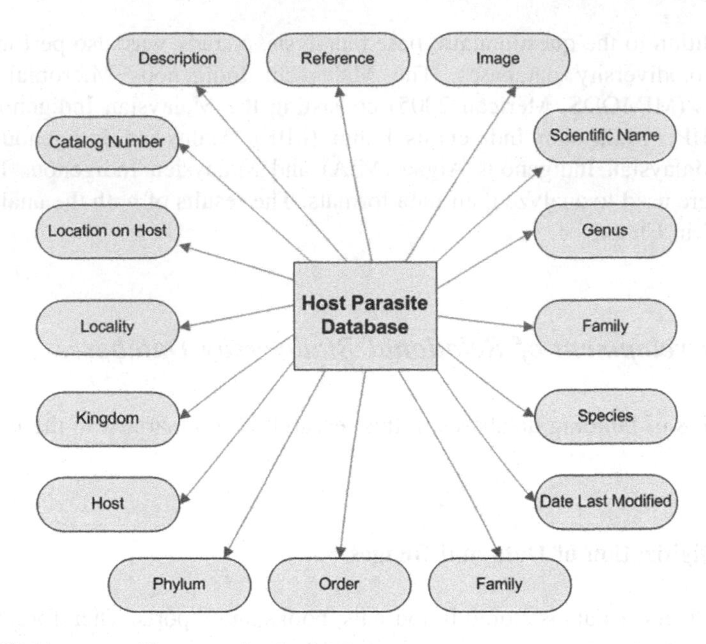

Fig. 4.2 Field names of host-parasite database

4.2.2.4 Selecting DBMS

In this research, heterogeneous DBMSs were selected to build biodiversity databases. The aim was to link these databases despite their heterogeneities. The selection of DBMSs was based on their usefulness, diversity, simplicity and complexity (see Table 3.2 in Chapter 3 for more details on DBMSs). The DBMSs used to build relational biodiversity databases in this research were DB2 by IBM, FileMaker Pro by Filemaker, Inc. and MySQL.

4.2.2.5 Implementing Tables and Relationships in DBMS

At this stage, the designs produced until Sect. 4.2.2.4 above were transformed into physical model in the DBMS. According to the DBMS notations, tables and relationships were implemented when building the databases.

Each database developed in this research contains tables which are related to one another.

The steps in Sect. 4.2.2.1 until Sect. 4.2.2.5 were implemented to build relational biodiversity databases in this research. These databases were used to test the database integration system proposed in this book (see Chapter 6). The results of these relational biodiversity databases are presented in Chapter 5.

In the next section, the methodology for the development of CABIR is explained.

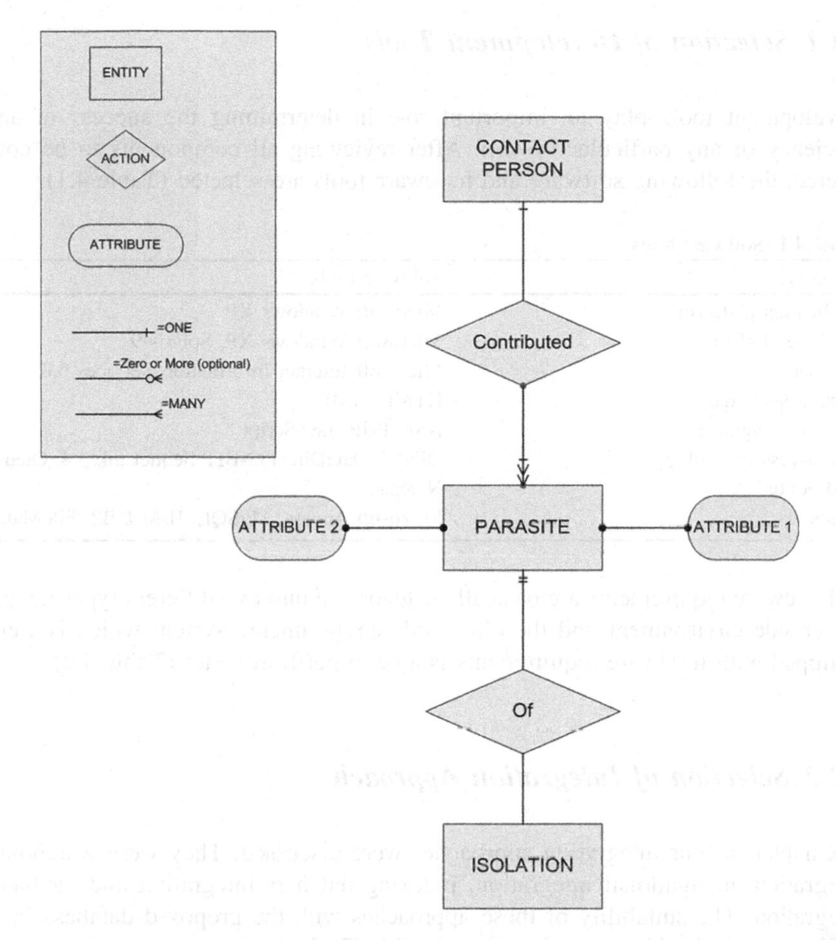

Fig. 4.3 E-R diagram for Host-Parasite database. *Diamond shape* represents the relationship between the tables, *rectangle* represents the tables in the database and *oval* represents the attributes in the table

4.3 Methodology for Building the Proposed Database Integration System

The first step towards building a new database integration system (CABIR, see Chapter 6 for details) in this research was to identify the software and hardware tools needed to build the system. Hence after, integration approach which best suites the requirements of the database integration system was selected. This was followed by selection of relational biodiversity databases which are the materials to test the system. In the final step, the database integration system was built. It involved integrating the databases, writing the xml wrappers and developing the query based portal (see Fig. 4.1).

4.3.1 Selection of Development Tools

Development tools play an important role in determining the success of and efficiency of any particular system. After reviewing all components to be considered, the following software and hardware tools are selected (Table 4.1):

Table 4.1 Software tools

Category	Software tools
Application platform	Microsoft Windows XP
Database platform	Microsoft Windows XP, Solaris 9
Web server	Microsoft Internet Information Services 6.0
Markup language	HTML, XML
Scripting language	ASP, PHP, JavaScript
Data access technology	ODBC, DataDirect32-BIT SequeLink 5.4, Oledb
Wed editor	Notepad
Databases	Microsoft Access, MySQL, IBM DB2, FileMaker

Hardware requirements are basically categorized into two different types i.e. the server-side environment and the client-side environment. System which is well-equipped with hardware requirements is able to perform better (Table 4.2).

4.3.2 Selection of Integration Approach

In Chapter 3, four integration approaches were discussed. They were warehouse integration, navigational integration, indexing flat files integration and mediator integration. The suitability of these approaches with the proposed database integration system in this research is presented in Table 4.3.

From the four integration approaches discussed, mediator based approach is the best method for CABIR. This is because the aim of CABIR to integrate remote and

Table 4.2 Hardware tools

(a) Server-side environment hardware tools (minimum requirement)	
Category	Hardware tools
Processor	Pentium III and above
RAM	128 Megabytes or higher
Hard disk space	At least 2 Gigabytes
Internet	1 Ethernet NIC connection
(b) Client-side environment hardware tools (minimum requirement)	
Category	Hardware tools
Processor	Pentium II and above
RAM	32 Megabytes or higher
Hard disk space	At least 1 Gigabytes
Internet	36.6 kbps connection

Table 4.3 Suitability of warehouse, navigational, indexing flat files and mediator approaches with database integration system proposed in this research

Integration approaches	Attributes	CABIR
Warehouse	Data from multiple sources stored in local warehouse	There is no local warehouse
	Generally not possible to integrate new databases without changing the schema of data warehouse	New databases will be integrated in the future
	Creates redundancy as there are records in the distributed database as well as in the local database therefore it is time consuming	Eliminates redundancy for efficiency
	Best suited to integrate limited data sources with stable database schemas	Data sources are not limited. It is expected to grow with time
Navigational	Effectively allows the representation of cases where the page containing the desired information is only reachable through a particular navigation path across other pages (Friedman et al. 1999)	Information is stored in database management systems (DBMS) and they might not have their own Web server
Indexing flat files	Databases to be integrated are provided as flat files	Databases to be integrated are provided in database management systems (DBMS)
	Does not provide a mechanism to integrate in-house relational databases, nor does it provide a mechanism to perform data cleaning and transformation for complex data mining	A mechanism to relate in-house rclational databases is required
	Not suitable for large amount of data	Expected to integrate large amount of data
Mediator	Concentrates on query translation	Concentrates on query translation
	Data sources can be distributed and maintained by the respective owner	Data sources are distributed and maintained by the respective owner

distributed databases matches the attributes of mediator based integration approach (see Table 4.3). Mediator based approach is also one of the characteristics of many of the current database integration systems (see Table 3.1 in Chapter 3).

4.3.3 Selection of Relational Biodiversity Databases for CABIR

The relational biodiversity databases used for the proposed database integration system in this research are existing databases as well as databases developed in Sect. 4.2.2.

Existing relational biodiversity databases are (Merican 2005);

- Malaysian Indigenous Algae (MIA)—Microsoft Access
- Malaysian Indigenous Bacteria (MIB)—Microsoft Access
- Malaysian Indigenous Virus (MIV)—Microsoft Access
- Malaysian Indigenous Protozoa (MIP)—Microsoft Access
- Malaysian Indigenous Fungi (MIF)—Microsoft Access
- Bactiden—MySQL

The above databases can be found online (Merican 2005). The new biodiversity databases used for the proposed database integration system are presented in Chapter 5.

4.3.4 Development of CABIR

The methodology involved in development of CABIR comprises of three components, which are described in the sub-sections below.

4.3.4.1 Database Integration (Providers) and Query Processing

According to the literature survey done in Chapter 3, building a database integration system initially necessitates integration of databases with the World Wide Web. It requires a Web server, application program and connectivity. In this research, Apache and Internet Information Service (IIS) were employed as Web servers for the databases to be Web accessible. ASP and PHP were chosen as scripting languages to integrate database with the Web. ASP was used for Windows-based providers whereas PHP for UNIX-based providers.

As for the database connectivity, ODBC, DataDirect32-BIT SequeLink 5.4 and Oledb were the driver managers used. Two methods were used to connect to databases in the proposed system. They are DSN and DSN-less connections. DSN here refers to the Data Source Name. Structured Query Language (SQL) was used to retrieve and manipulate the data from the relational databases.

Each database in the proposed system requires a provider which will connect the database to the Web server. The provider also contains the SQL strings to perform the desired query to the database according to the request from the wrappers (see Sect. 4.3.4.2 on wrappers). The data extracted from the database was returned as XML schema and XML document. The processing of these documents is explained in Sect. 4.3.4.2. Figure 4.4 shows the architectural view of the database integration process for one database.

In Fig. 4.4, the application sent a query to the Web server, through the Internet. The query was then forwarded to the specific provider. The provider returned results in XML format which was then converted into readable HTML format.

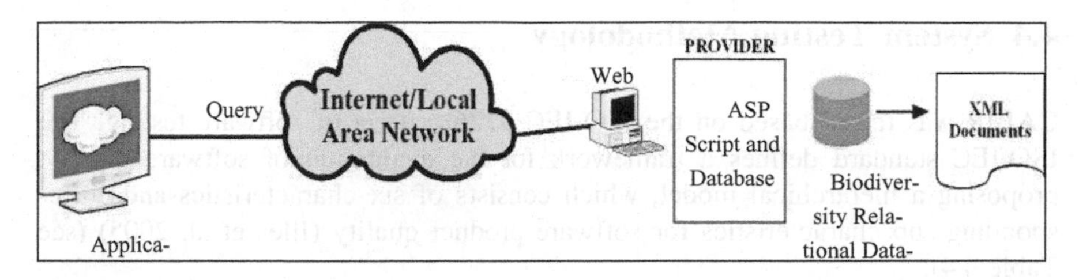

Fig. 4.4 Architectural view of database integration proposed in this research (In this diagram, only one database is used)

Figure 3.5 illustrated the process for a single database. The Web server contains a connection with the back-end database. The document that contains the ASP/PHP script and database connection is called a Provider. In this research, heterogeneous databases are used and they are queried simultaneously.

4.3.4.2 XML: Wrappers and Validation

XML is used extensively in the development of CABIR. The importance and use of XML was described in Sect. 3.4.4, Chapter 3. In CABIR, the results returned by the query are stored in the form of XML documents. Results searched from each database will be put in an XML document. This XML document may contain some errors and unstructured data format. Therefore, it will then go through a validator to map against the XML schema. The XML results after the validation will be transformed into HTML using the Extensible Stylesheet Language Transformations (XSLT). Therefore, XML plays are very important part in the formation of CABIR.

4.3.4.3 Search Engine and User Interface

Another important step for building the database integration system in this research was building the application server to interact with the users. There are two levels forming the application server which are the search engine and user interface.

The search engine was developed using the ASP, HTML and XML. The search text box which was build using a HTML form accepts a query in the form of texts and sends it to the destination(s) or database(s) selected by user via TCP/IP and XMLHTTP methods. The Uniform Resource Locator or IP address provided in the application will determine the location of the database(s).

CABIR's interface was developed using Hypertext Markup Language (HTML), VBScript, JavaScript and Active server pages. It contained a form with a search textbox and options to select the data providers (databases), besides the appropriate background and images to enhance the appearance of the main page.

4.4 System Testing Methodology

CABIR was tested based on the ISO-IEC 9126 criteria of software testing. The ISO/IEC standard defines a framework for the evaluation of software quality, proposing a hierarchical model, which consists of six characteristics and corresponding sub characteristics for software product quality (Illes et al. 2005) (see Table 4.4).

Functionality and Usability are tested using a heuristic checklist which was given to an expert. The checklist is presented in Table 4.5.

The reliability and efficiency test was conducted using a performance test method following a standard model used by Roderic (2005). For each database in Table 4.6, the species name was searched 100 times and response time was recorded from the time the query was made until the time the results were returned. The results of the performance benchmarks are shown in Table 6.10 and illustrated in Fig. 6.30 in Chapter 6.

As for the maintainability and portability of CABIR, it was identified by looking at the underlying architecture of CABIR. The results of the test are presented in Chapter 6 of this book.

Table 4.4 Six Characteristics for Software Quality following the ISO-IEC 9126 Standard

Main characteristic	Description	Sub-characteristics
Functionality	A set of attributes that bear on the existence of a set of functions and their specified properties. The functions are those that satisfy stated or implied needs.	Suitability, accurateness, interoperability, compliance, security
Reliability	A set of attributes that bear on the capability of software to maintain its level of performance under stated conditions for a stated period of time.	Maturity, fault tolerance, recoverability
Usability	A set of attributes that bear on the effort needed for use, and on the individual assessment of such use, by a stated or implied set of users	Understandability, learnability, operability
Efficiency	A set of attributes that bear on the relationship between the level of performance of the software and the amount of resources used, under stated conditions	Time behavior, resource behavior
Maintainability	A set of attributes that bear on the effort needed to make specified modifications	Analyzability, changeability, stability, testability
Portability	A set of attributes that bear on the ability of software to be transferred from one environment to another	Adaptability, installability, conformance, replaceability

Table 4.5 Functionality and Usability checklist

Criteria	Yes	No
Functionality		
Link remote databases in a networked environment		
Support heterogeneous data format		
Support heterogeneous database management systems		
Link databases hosted in Windows and UNIX based platforms		
Provide data security for database owners by allowing them to keep and maintain their own data and to choose information to be shared and linked		
Usability		
It is easy to know the current location within the overall map of the site		
It is clear what information is available at the current location		
It is clear where you can go from the current location		
It is always easy to return to the Home Page		
Links are used and appear in standard web style		
Menus are used and appear in standard web style		
Link labels match destination page titles or headers		
Labels and links are described clearly		
Color choices allow for easy readability		
If necessary, error messages are clear and in plain language		
The content language is clear and simple		

Table 4.6 Relational biodiversity databases tested with CABIR

Name	DBMS	Operating systems	No of records	Location
Malaysian Indigenous Algae	Microsoft access	Windows XP	7	Bioinformatics research Lab, University Malaya
Malaysian indigenous fern	FileMaker	Windows XP	125	Zoology Museum, UM
Photogallery	MySQL	Windows XP	48	Bioinformatics Research Lab, University Malaya
South East Asia Fruit Flies	MySQL	Solaris 9	10	Zoology Museum, UM
Biodiversity sample database	MySQL	Linux	10	MIMOS, Technology Park Malaysia
Virus sample database	DB2	Windows XP	21	Bioinformatics research Lab, University Malaya

4.5 Conclusion

In this chapter, methods used for the development of a database integration system were described. The selection of tools and methods were done following the literature review presented in Chapter 3. In the following chapters, the results obtained are presented. Chapter 4 presents and discusses the results of data format and relational biodiversity databases developed in this chapter. Chapter 6 presents and discusses the results of the database integration system developed using methods and materials presented in this chapter. Chapter 6 also demonstrates the results of the system testing done using methods presented in this chapter.

References

Friedman, M., Levy, A., & Millstein, T. (1999). Navigational plans for data integration. In *Proceedings of the National Conference on Artificial Intelligence (AAAI)*, pp 67–73.

Illes, T., Herrmann, A., Paech, B., & Rückert, J. (2005). criteria for software testing tool evaluation—a task oriented view. In *Proceedings of the 3rd World Congress for Software Quality,* Munich, Germany.

Merican, A. F. (2005). *Merican Research Lab* (online). University Malaya, Kuala Lumpur. http://amirmericanlab.um.edu.my/ (Accessed 12 Aug 2005).

Post, G. V. (2005). *Database management systems. Designing and building business applications*. New York: McGraw Hill.

Roderic, D. M. (2005). A taxonomic search engine: Federating taxonomic databases using web services. *BMC Bioinformatics, 6*, 48.

Xia, Y., Stinner, R. E., & Chu, P. C. (2002). Database Integration with the Web for biologist to share data and information. *EJB Electronical Journal of Biotechnology* ISSN: 0717-3458.

Chapter 5
Biodiversity Databases

5.1 Introduction

Biodiversity data format and relational databases were built and described in Sect. 4.2. The biodiversity data format is presented and explained in Sect. 4.2.1 whereas the relational biodiversity databases are presented in Sect. 4.2.2. Besides the existing databases, the databases described in this section are used as data sources to test the proposed database integration solution in this research.

5.2 Biodiversity Data Format

Data was gathered from researchers in various biodiversity fields as well as from existing biodiversity databases. A data format was developed based on the analysis in Sect. 4.2.1. This format was used as a standard to name fields in the new biodiversity databases built in this research. Although most of the data fields in this data format originated from Darwin Core V2 format, new field additions were included (see Table 5.1).

The data format presented in Table 5.1 can be altered or added according to the requirements of a database owner. It is important to note that the key fields in this format is inline a global standard which is Darwin Core V2, therefore databases built using this format can be shared with other biodiversity databases in the world.

5.3 Relational Biodiversity Databases

The methodology used to develop biodiversity database in Chapter 4 are summarized below:

(1) Digitization of data and images from manuals
(2) Naming the database fields using a standard naming convention or data format

S. K. Dhillon and A. S. Sidhu, *Data Intensive Computing for Biodiversity*, Data, Semantics and Cloud Computing 485, DOI: 10.1007/978-3-642-38047-1_5,

Table 5.1 Data format for biodiversity databases. New additions are highlighted with darkened background

Field	Data type	Description
Date last modified	Text	ISO 8601 compliant stamp indicating the date and time in UTC(GMT) when the record was last modified. Example: the instant "November 5, 1994, 8:15:30 am, US Eastern Standard Time" would be represented as "1994-11-05T13:15:30Z"
Collection code	Text	A unique alphanumeric value which identifies the collection within the institution
Institution code	Text	A "standard" code identifier that identifies the institution to which the collection belongs. No global registry exists for assigning institutional codes. Use the code that is "standard" in your discipline
Catalog number	Tex	A unique alphanumeric value which identifies an individual record within the collection. It is recommended that this value provides a key by which the actual specimen can be identified. If the specimen has several items such as various types of preparation, this value should identify the individual component of the specimen
Scientific name	Text	The full name of lowest level taxon the Cataloged Item can be identified as a member of; includes genus name, specific epithet, and sub specific epithet (zool.) or infraspecific rank abbreviation, and infraspecific epithet (bot.) use name of suprageneric taxon (e.g., family name) if cataloged item cannot be identified to genus, species, or infraspecific taxon
Kingdom	Text	The kingdom to which the organism belongs
Class	Text	The class name of the organism
Order	Text	The order name of the organism
Family	Text	The family name of the organism
Genus	Text	The genus name of the organism
Species	Text	The specific epithet of the organism
Morphology	Text	Form or shape of an organism
Colour	Text	The perception of the frequency (or wavelength) of light
Advantages	Text	The usage of the organism
Characteristics	Text	The habitat and culture which grown by the organism
Pathogenesis	Text	The mechanism by which a certain etiological factor causes disease (*pathos* = disease, *genesis* = development)
Pathogenicity	Text	The ability of an organism to cause disease in another organism
Reference	Text	A note in a publication referring the reader to another passage or source
Contact person	Text	The name(s) of the contact person(s) responsible for collection the specimen or taking the observation
source of isolation	Text	Location where carry out the isolation of organism
Cell morphology	Text	Form or shape of the cell
Colony morphology	Text	Form or shape of the colony

(continued)

Table 5.1 (continued)

Field	Data type	Description
Email	Text	The contact of the collector(s) responsible for collection the specimen or taking the observation
Metabolism	Text	The biochemical modification of chemical compounds in living organisms and cells
Group name	Text	The group name of the organism
Organism	Text	The complex adaptive system of organs that influence each other in such a way that they function
Type of nucleic acid	Text	The description of nucleic acid. Single-stranded or double-stranded of DNA or RNA
Geographical distribution	Text	Location that can get the organism
Transmission	Text	The passing of a disease
Host range	Text	The collection of hosts that an organism can utilize as a partner
Disease symptoms	Text	Sign of the existence of a condition
Diagnosis	Text	Identification of a disease or condition after observing its signs
Phylum	Text	The phylum (or division) to which the organism belongs
Hosts	Text	An organism that harbors a virus, parasite, mutual partner, or commensally partner
Source	Text	Person or book etc. supplying information
Growth	Text	Place where the organism grow
Life	Text	A period involving one generation of an organism
Size	Text	The measurement of organism
Treatment	Text	Therapy used
Description	Text	Definition of the organism
Locality	Text	The locality description (place name plus optionally a displacement from the place name) from which the specimen was collected. Where a displacement from a location is provided, it should be in un-projected units of measurement
Synonyms	Text	Similar or identical meanings of organism and are interchangeable
Authority	Text	The legitimacy, justification and right of the organism
Image	Text	Link to the image
Gram	Text	A type of staining to identified organism that can divided to gram negative and gram positive
Width	Text	A measurement of the distance from the right to the left of organism
Height	Text	A measurement of the distance from the bottom to the top of organism
Title	Text	A prefix or suffix added to a organism's name to signify
Caption	Text	A concise and descriptive bit of text that labels an organism
Website	Text	The Uniform Resource Locator (URL) which can link to the organism

(3) Designing tables and relationships among the entities
(4) Choosing an appropriate DBMS
(5) Implementing tables and relationships in DBMS.

Table 5.2 Relational databases developed in this research. The table shows the database names, DBMS used, platform on which they were built

Database name	DBMS	Operating systems	No. of records	Location
South East Asia fruit flies database	MySQL	Solaris 9	100	Zoology museum, University Malaya
Host parasite database	FileMaker	Windows	150	Zoology museum, University Malaya
Fern database	FileMaker	Windows	125	Zoology museum, University Malaya
Zingiberaceae database	FileMaker	Windows	140	Zoology museum, University Malaya

These five steps are important to be followed in their specified order for the success of a database. Using these methods (see Sect. 4.3, Chapter 4 for more details), four databases were built. These databases contain primary data sources collected from local researchers. A summary is presented in Table 5.2. Overview and designs of these databases are presented in the following subsections.

5.3.1 South East Asia Fruit Flies Database

The South East Asia Fruit Flies database was developed using primary data sources of Dr Chua Tock Hing from Monash University, Malaysia. South East Asia Fruit Flies database is hosted on a Solaris platform. There are five tables defined in this database: data_sp, data_bib, data_host, Host_Photo and Flies_photo (see Fig. 5.1a). The screen design of South East Asia Fruit Flies database is illustrated in Fig. 5.1b.

5.3.2 Host-Parasite Database

There are eighteen tables defined in the Host-Parasite database which are Parasite, Host, Genus, Phylum, Class, Order, Family, EditPage, GenusEdit, PhylumEdit, ClassEdit, OrderEdit, FamilyEdit, HostEdit, GenusTable, FamilyTable, HostTable and SearchPage. Figure 5.2 shows the relationship between tables while Fig. 5.3 shows the screen designs of Host-Parasite database, which consists of a search page, family list, editing page and parasite detail.

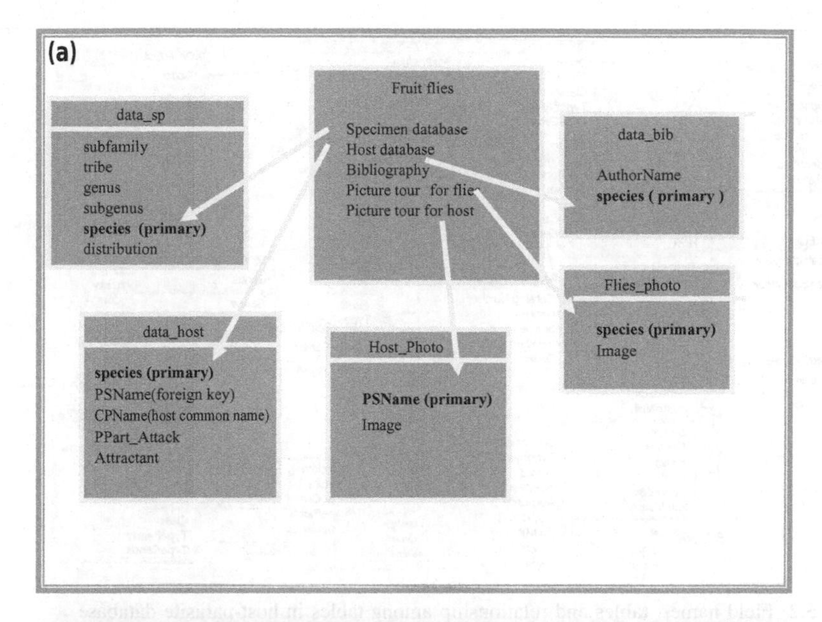

Fig. 5.1 **a** Field names, tables and relationship among tables in South East Asia fruit flies database. **b** Screen design of South East Asia fruit flies database

5.3.3 Malaysian Indigenous Fern Database

There are three tables defined in this database which are Taxonomy, Classification, and Image. Figure 5.4 shows relationships between these tables. The Malaysian Indigenous Fern database requires authentication. A user will need to key in correct username and password to log on to the database. Once logged on,

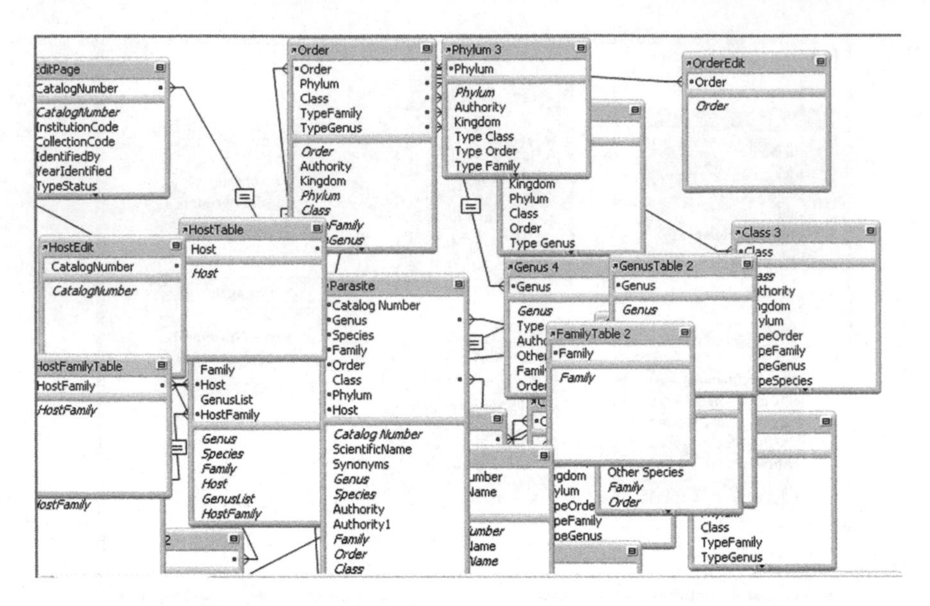

Fig. 5.2 Field names, tables and relationship among tables in host-parasite database

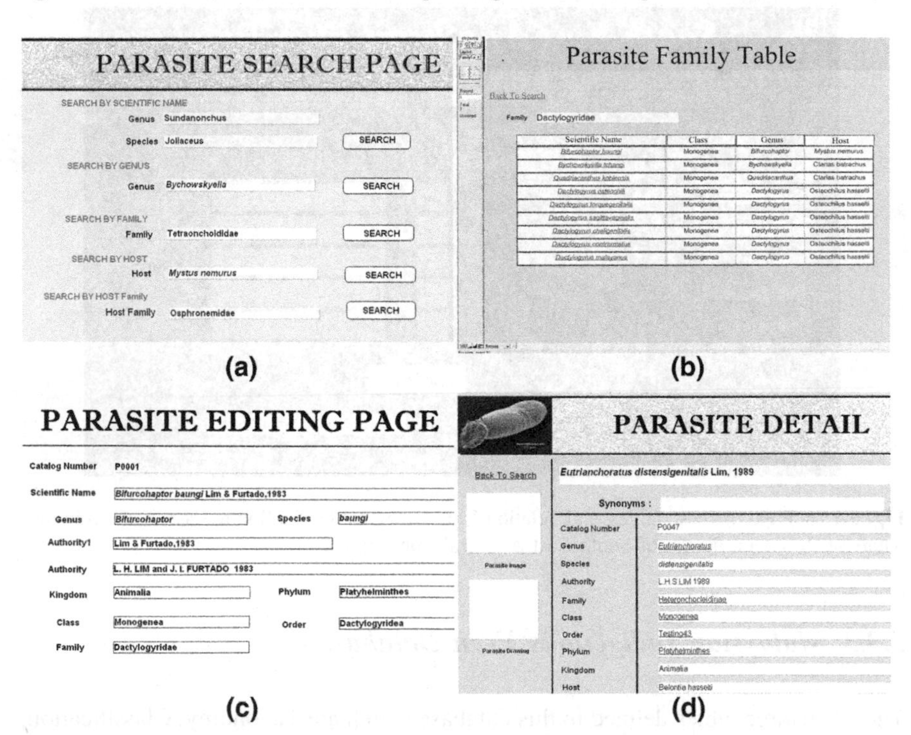

Fig. 5.3 Screen designs of host-parasite database. **a** This screen shows the search page for host parasite database. Search can be done through scientific name, genus, family, host and host family. **b** This screen shows the results in tabular form. **c** Host Parasite data can be edited through the parasite editing page. **d** This screen shows the details of parasite data

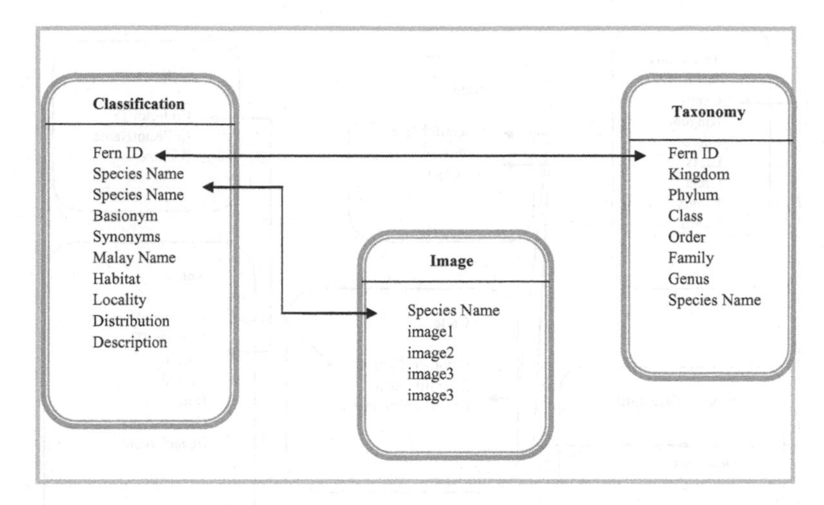

Fig. 5.4 Field names and relationship among tables in Malaysian indigenous fern database

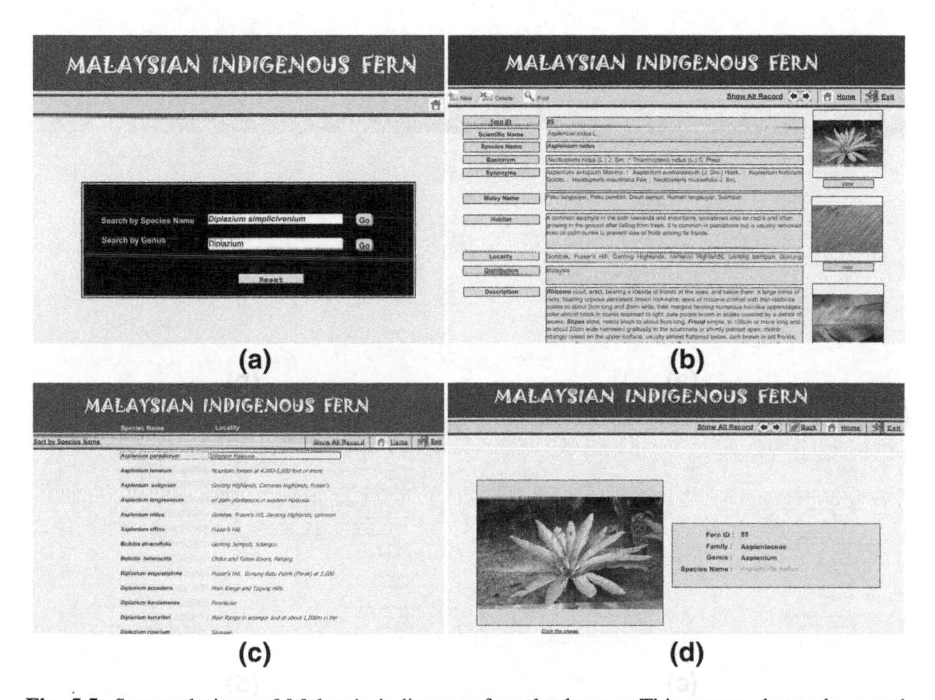

Fig. 5.5 Screen designs of Malaysia indigenous fern database. **a** This screen shows the search page. Data can be searched through the species name or genus name. **b** This screen shows the add record page. New records can be added by filling in the details in the form. **c** Records can be edited through the edit record page presented in this screen. **d** This screen shows the fern details

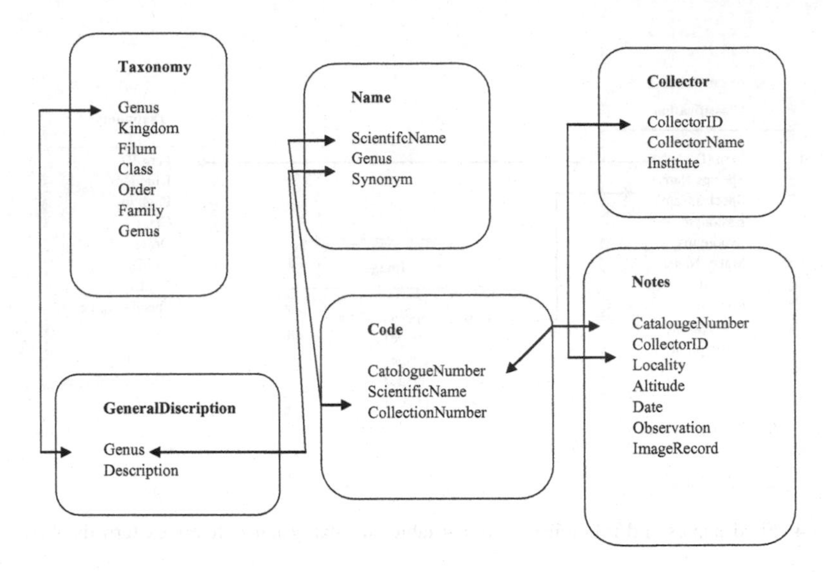

Fig. 5.6 Field names and relationship among tables in the herbarium of Zingiberaceae database

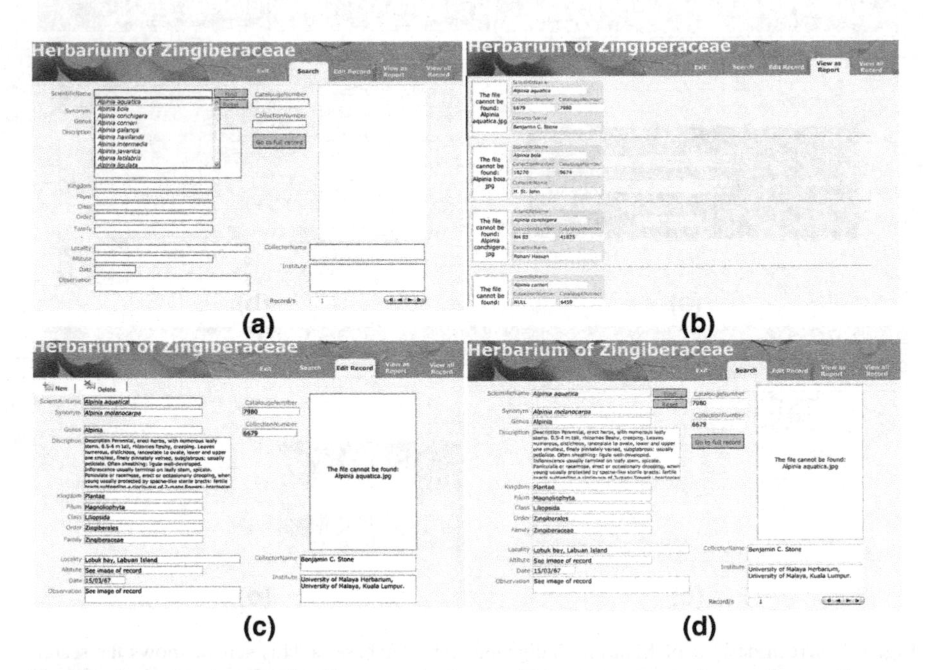

Fig. 5.7 Screen designs of Zingiberaceae database. **a** This screen shows the search page. Data on Zingiberaceae can be searched through scientific name. **b** This screen shows the results of search. **c** Records can be edited through the edit page. **d** The details of search is presented in this screen. There is a button to go to full record

information on ferns can be searched by Species or Genus names. Figure 5.5 shows the screen designs of the Malaysian Indigenous Fern database, which consists of a search page, add records page, records edit page and details page.

5.3.4 Herbarium of Zingiberaceae Database

There are six tables defined in this database which are Taxonomy, Code, Collector, GeneralDescription, Notes and Name. Figure 5.6 shows the relationships between these tables. The Herbarium of Zingiberaceae database requires authentication of username and password. Once logged on, information on zingiberaceaes can be searched by Scientific Name. Figure 5.7 shows the screen designs of Herbarium of Zingiberaceae database, which consists of a search page, results page, edit records page and search details.

5.4 Conclusion

The relational biodiversity databases that were built and described in this chapter were used to test CABIR. These databases vary in their DBMSs, operating systems and type of biodiversity data (see Table 5.2). The use of these databases is further explained in Chapter 6.

Chapter 6
Proposed Solution

6.1 Introduction

As presented and discussed in Chapter 2, DiGIR implementation on MIMODS was unsuccessful and time consuming. In addition, it has limitations such as reading from FileMaker DBMSs, retrieving images and a rigid data format. Therefore an alternative database integration system was developed to retrieve data from remote heterogeneous relational biodiversity databases (see Sect. 4.3).

The methods used to develop CABIR were described in Chapter 4 (see Sect. 4.3). CABIR is not only simple but solved some of the problems faced in DiGIR (see Chapter 5 for DiGIR's results). The CABIR system was tested with biodiversity data sources presented in Table 6.1. These databases comprise of existing biodiversity databases and new biodiversity databases presented in Chapter 5.

6.2 Hierarchical View of CABIR

The hierarchical design and the work flow of the entire system are presented in Fig. 6.1a, b.

6.3 Proposed Database Integration System: CABIR

6.3.1 System Architecture

The architecture in Fig. 6.2 shows the essential components which are *Presentation Layer* which consists of the query based portal, *Resource Locator*, *XML wrapper*, *Database Connectivity*, *Data Filtering and Data Cleaning* and *Providers.*

S. K. Dhillon and A. S. Sidhu, *Data Intensive Computing for Biodiversity*,
Data, Semantics and Cloud Computing 485, DOI: 10.1007/978-3-642-38047-1_6,

Table 6.1 Biodiversity data sources used for CABIR

DBMS	Name
Microsoft Access	Malaysian Indigenous Algae
Microsoft Access	Malaysian Indigenous Bacteria
Microsoft Access	Malaysian Indigenous Virus
Microsoft Access	Malaysian Indigenous Protozoa
Microsoft Access	Malaysian Indigenous Fungi
FileMaker	Host parasite
FileMaker	Fern
FileMaker	Zinger
MySQL	Bactiden
MySQL	Photogallery
MySQL	South East Asia fruit flies
MySQL	Sample database (biodiversity)
DB2	Sample database (virus)

Based on the user's selection, the resource locator will look up for the resources which are web biodiversity data sources connected using CABIR. At this point users can select more than one database and CABIR will execute the processing simultaneously to these repositories. Once the resource locator has identified the web data sources, query will be sent via the TCP/IP protocol using the uniform resource locator (URL) to retrieve all the necessary data. XML wrappers contains provider information document, xml schema and xml documents containing query results. The provider information document consists of the database connectivity details, namespace and query statements. This document will be installed at the client side where the database resides. The xml schema will map the query results into a well formed data structure which applies the Darwin Core V2 global standard (Vieglais 2003). Once the results are produced in the XML document, the data goes through a clean-up phase to meet the user's requirements. At this point, surplus data and empty fields are filtered out. Thus, data is sent to the presentation layer, which is then converted into HTML for viewing.

6.3.1.1 Presentation Layer

There are two levels forming the presentation layer. They are the user interface and search engine (Fig. 6.3).The interface of CABIR was developed using Hypertext Markup Language (HTML), VBScript, JavaScript and Active server pages. It contained a form with a search textbox and options to select the data providers (databases).

The database integration system is a component of the Living Web portal (Sarinder et al. 2006) which was developed prior to this research. It contains a link to all the databases under the Integrated Biological Database Initiative (IBDI) (Sarinder et al. 2005) and also a link to the data integration system built in this research.

(a)

Fig. 6.1 **a** Hierarchical view of database integration system. The provider and XML wrapper (highlighted box) reside in the remote machine where the database is stored. The query based portal, resource and results reside in the application machine. Users will only see the search page as the rest are hidden from users **b** Flow of CABIR

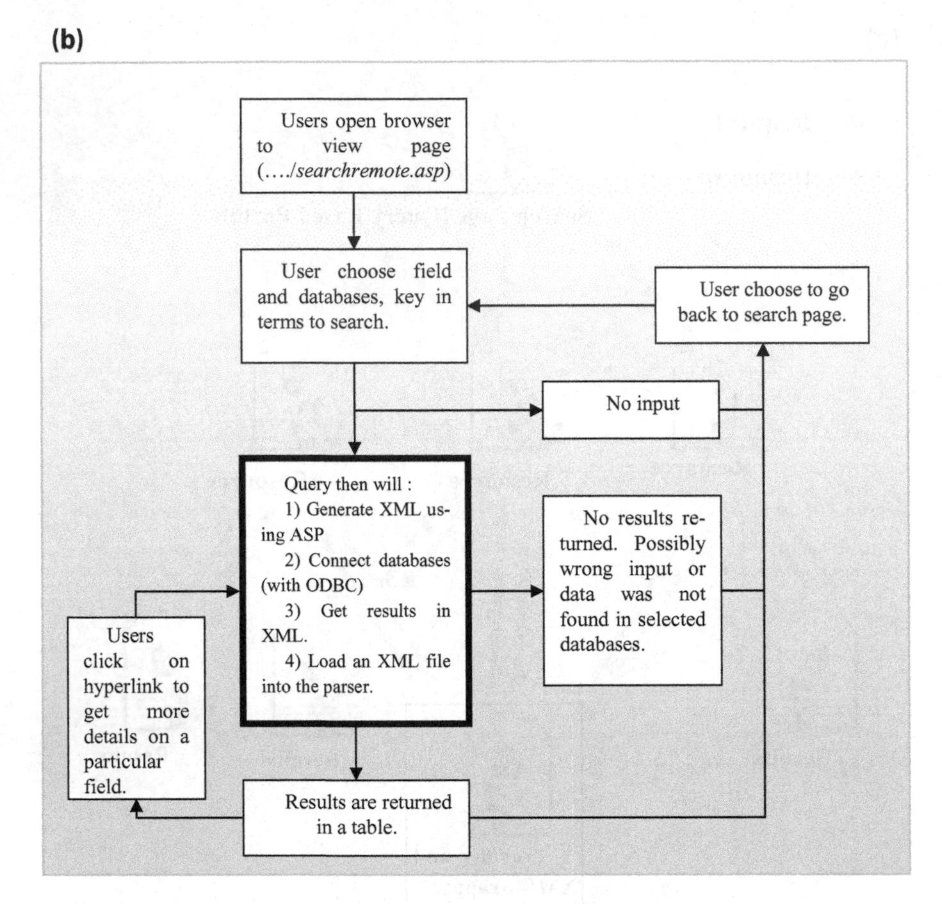

Fig. 6.1 (continued)

The search engine is responsible for sending the query to resource locator using the HTTP POST method. [using the library in article (West 2001)]. The engine was implemented using ASP scripting language and HTML.

6.3.1.2 Resource Locator

The main duty of the resource locator is to send the query to the exact destination(s). The search engine sends the query string supplied by user to the resource locator which then constructs a uniform resource locator (URL) for the corresponding database. In simpler words, the resource locator here performs the intermediary duties between the search engine and the providers. The query is sent to the designated providers using a search algorithm via XML and HTTP methods. Data from the providers are sent back using the GET method.

An instance of the XML HTTP open method is presented in Fig. 6.4.

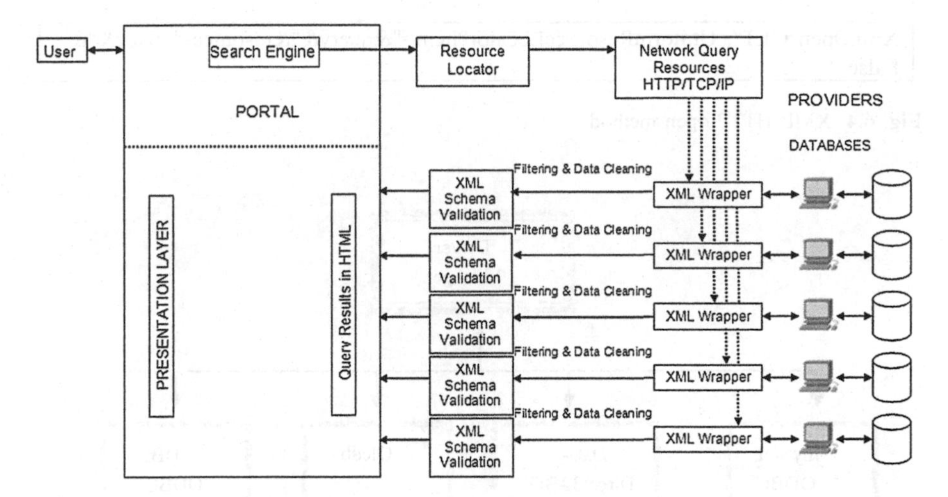

Fig. 6.2 CABIR Architecture

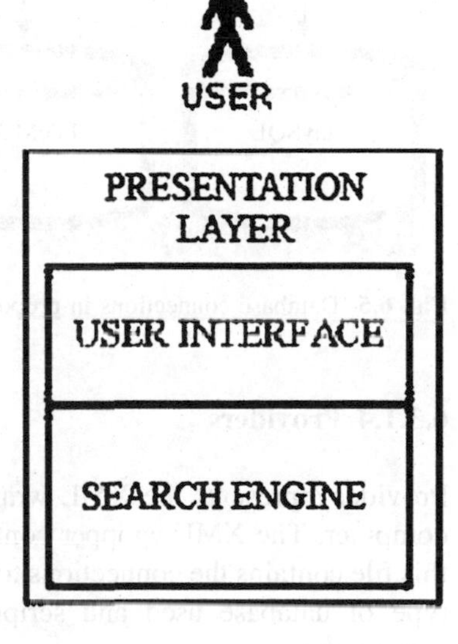

Fig. 6.3 Architectural view of presentation layer

6.3.1.3 Network Query Resources (TCP/IP)

The CABIR application runs on the network. Therefore the network protocols are core components needed to establish a link between the portal and remote databases. The resource locator sends the query to the providers through the TCP/IP protocol. The operation of TCP/IP protocol is not discussed thoroughly here as it a standard protocol used a Web service. (see Sect. 3.4.7).

```
Xml.Open"GET","UniformResourceLocator?item="&query&"&resource=" &arr&"&",
False
```

Fig. 6.4 XML HTTP open method

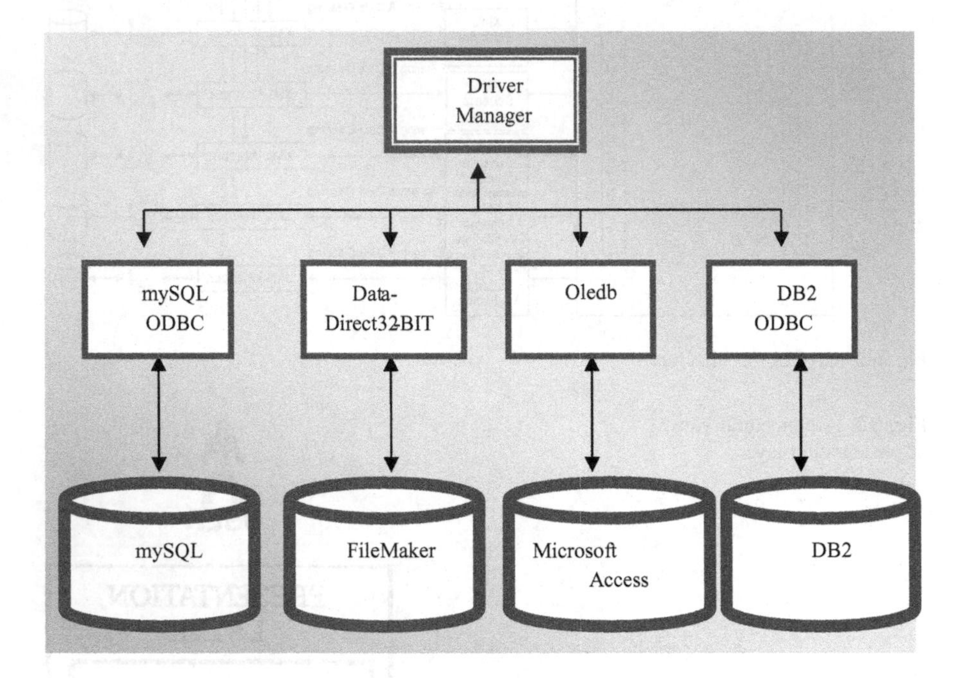

Fig. 6.5 Database connections in proposed database integration

6.3.1.4 Providers

Provider represents the XML wrapper as well as the database(s) that reside in a computer. The XML wrapper contains a file which is called the provider file and this file contains the connections to the database(s). The connection depends on the type of database used and scripting language used to communicate with the database(s).

For the development of CABIR, ODBC, DataDirect32-BIT SequeLink 5.4 and Oledb were the driver managers used (see Fig. 6.5). Two methods were used to connect to databases in the proposed system. They are DSN and DSN-less connections. DSN here refers to the Data Source Name. Figures 6.6, 6.7, 6.8, 6.9, 6.10 and 6.11 show the DSN-less and DSN connection with Microsoft Access, FileMaker, MySQL and DB2 DBMSs.

Structured Query Language (SQL) was used to retrieve and manipulate the data from the relational databases. It was also applied to relate the tables using the *JOIN* command.

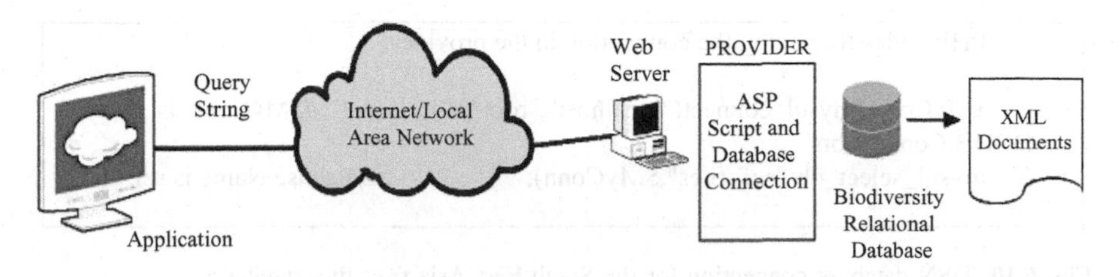

Fig. 6.6 Architectural view of database integration proposed in this research (In this diagram, only one database is used)

```
ASP codes for setting the connection in the provider

If strDatabaseType = "Access" Then                        # check database type
strConn = "Provider = Microsoft.Jet.OLEDB.4.0;" & _       # provider type
"Data Source =directory_of_database_file;" & _            #setting the location
of database file
End If
```

Fig. 6.7 DSN-less database connection for the algae database. The database was built using Microsoft access in windows operating systems

```
ASP codes for setting the connection in the provider.

Set MyConn = Server.CreateObject("ADODB.Connection") #Creating an object
MyConn.Open "Dsn=HOZ;" & _              #DSN name =HOZ
"Database=HOZ.fp7;" & _                                  #Database name
"Uid=******; " & _                                         #User Name
"Pwd=*****"                                                 #Password
```

Fig. 6.8 DSN database connection for the Zingiberaceae database. The database was built using FileMaker in Windows operating systems. The DSN connection first requires a DSN name to be set in the computer where the database is located. The name is then denoted in the ASP codes

```
ASP codes for setting the connection in the provider.

Set MyConn = Server.CreateObject("ADODB.Connection")
MyConn.Open "Driver={MySQL}; Server=localhost; Database=bactiden; Us-
er=ODBC;Password=;Option=4"
```

Fig. 6.9 DSN database connection for the Bactiden. This database was built using mySQL in Windows operating systems. The DSN connection first requires a DSN name to be set in the computer where the database is located. The name is then denoted in the ASP codes

```
PHP codes for setting the connection in the provider.

$MyConn=mysql_connect("localhost","root","");          #$MyConn is of type
"ADODB.Connection"
mysql_select_db("sef_flies",$MyConn);          #Database Name is sef_flies
```

Fig. 6.10 DSN database connection for the South East Asia fruit flies database

```
ASP codes for setting the connection in the provider.

#$MyConn is of type "ADODB.Connection"
Set MyConn = Server.CreateObject("ADODB.Connection")

#Parameters of connection
myConn.Open          "DSN=DB2VIRUS;UID=user;
PWD=kesuma_215;DBQ=SAMPLEDB"
```

Fig. 6.11 DSN database connection for the virus database. This database was built using DB2 in Windows operating systems

The provider file also contains the SQL strings to perform the desired query to the database according to the request from the wrappers. (see Sect. 4.3.4 on wrappers). The data extracted from the database was returned in the form of an XML document.

This database was built using mySQL in Solaris 9 operating systems.

6.3.1.5 XML Schema Validation

The XML document which contains the results of the query in the XML format goes through a stage where data cleaning and filtering is performed. This step is called XML schema validation. The XML document is validated against a schema which was constructed specifically for CABIR. An instance of the schema is presented in Fig. 6.12. The validation between the XML document and XML schema is done by a validation file. This validation file checks for errors such as null entries, incorrect data type and structure.

Finally, the results after validation is passed back to the presentation layer where they are converted into HTML for viewing. This is done by using the XSLT (Extensible Stylesheet Language Transformations). XSLT is a transformation language for converting XML instances.

```
?xml version="1.0" encoding="iso-8859-1"?>
<xs:schema xmlns:xsi="http://www.w3.org/2001/XMLSchema-instance"
attributeFormDefault="unqualified"          elementFormDefault="qualified"
xmlns:xs="http://www.w3.org/2001/XMLSchema">
<xs:element name="records">
<xs:complexType>
<xs:sequence>
<xs:element name="record" maxOccurs="unbounded" >
<xs:complexType>
<xs:sequence>
<xs:element name="Number" type="xs:integer" />
<xs:element name="DateLastModified" type="xs:date"  nillable="true" />
<xs:element name="CollectionCode" type="xs:string" />
<xs:element name="InstitutionCode" type="xs:string" />
<xs:element name="CatalogNumber" type="xs:string" />
<xs:element name="ScientificName"  type="xs:string"/>
<xs:element name="Family" type="xs:string" />
<xs:element name="Genus" type="xs:string" />
<xs:element name="Morphology" type="xs:string" />
```

Fig. 6.12 Part of XML schema (XSD file)

6.3.2 System Features

CABIR has a simple and user friendly interface. Users can select fields to search from drop down menu (see Fig. 6.13), key in the search in the textbox (see Fig. 6.14) and select the database they want to query on (see Fig. 6.15). There is a help link that will take users to the help screen which will assist users on using the system (see Fig. 6.16). The results of search are presented in a manner that is easy to understand and has features that can assist a user such as timer (see Fig. 6.17) and total count of records found in each database (see Fig. 6.18). Besides that, users can jump to records in different databases by clicking on the hyperlinks in the results page (see Fig. 6.19).

The search results are displayed using alternate green and white. (see Fig. 6.20). This is to increase user readability. For each record, only Family, Genus and Species are displayed to avoid congestion in the screen. If users want more details of a specific record, they can navigate through the links. (see Fig. 6.21).

The features described here are important for users as it helps them navigate through the system smoothly. The features above make CABIR a simple and dynamic database integration system.

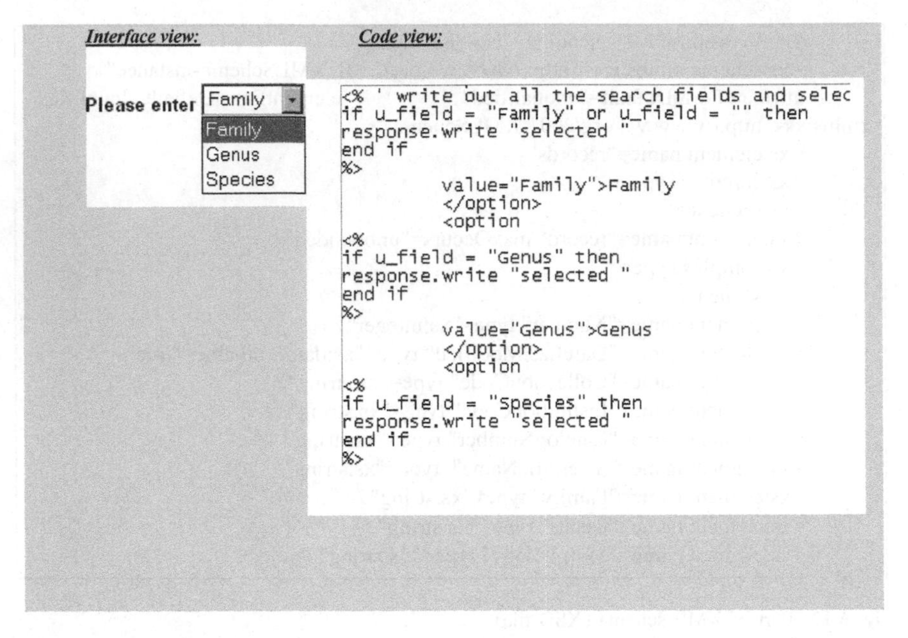

Fig. 6.13 Interface view and ASP codes to create an option to choose a field (Family, Genus or Species) to search on

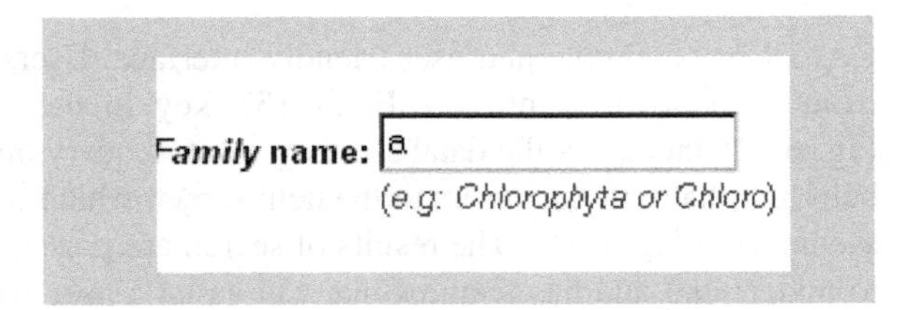

Fig. 6.14 Input typed in the text box will be captured as a query. After pressing "Submit", users will be redirected to the *Search Result* page

6.3.3 CABIR and FileMaker Databases

CABIR has demonstrated data retrieval from FileMaker DBMSs. Figures 6.22, 6.23 and 6.24 show the results from FileMaker DBMSs which are Host Parasite, Herbarium of Zingiberaceae and Malaysian Indigenous Fern. Results were obtained in 0.9 s, 0.63 s and 0.86 s respectively.

6.3.4 Image Retrieval

CABIR is able to retrieve images from remote relational databases (see Figs. 6.25, 6.26). User can retrieve images from Bactiden and Photogallery databases. The

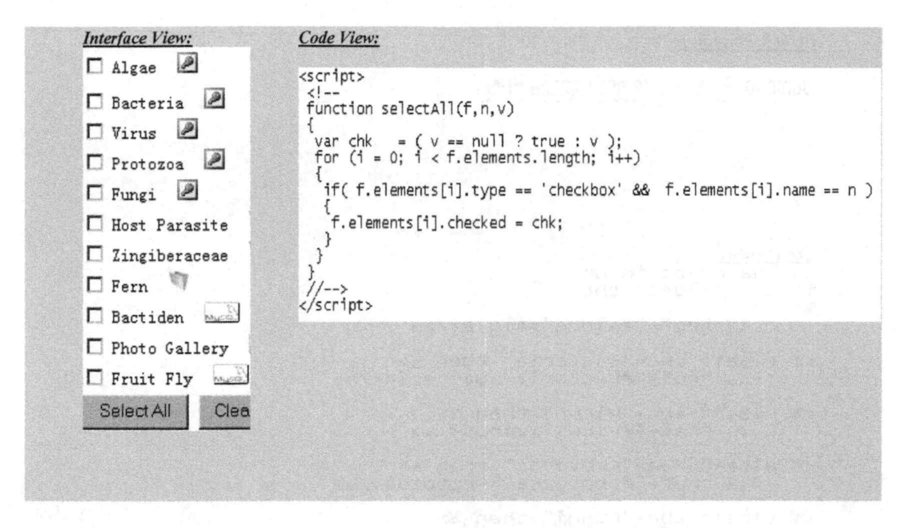

Fig. 6.15 Interface view and the Java Script code for database selection. User can choose by selecting one or more check boxes or by pressing the "Select All" button for all databases

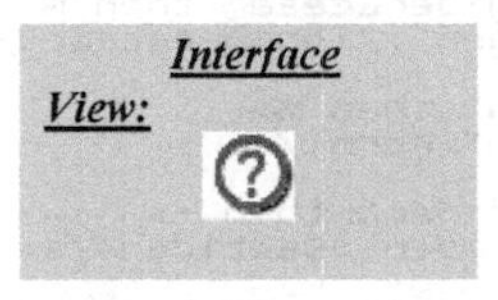

Fig. 6.16 Interface view for help link. When users click on the help link, they can view fields that can be searched for each database

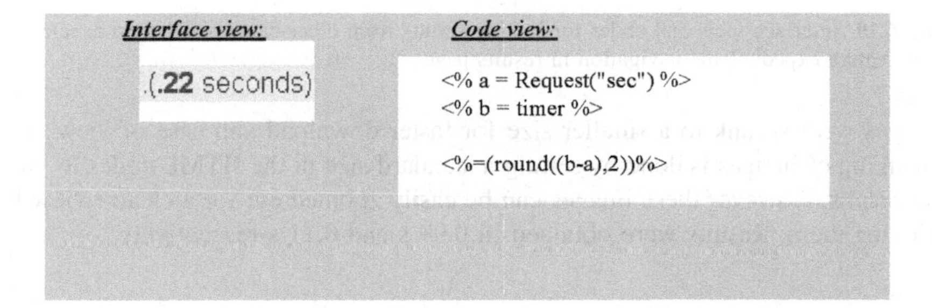

Fig. 6.17 Interface view and ASP codes for timer. Search timer is displayed in the results page

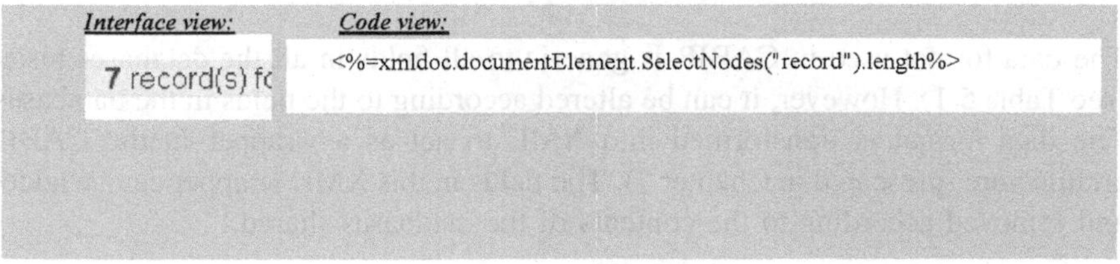

Interface view:

7 record(s) fc

Code view:

```
<%=xmldoc.documentElement.SelectNodes("record").length%>
```

Fig. 6.18 Total record(s) found is represented by the ASP code (count selectNodes method)

Interface view:

Jump to: Algae Bacteria Protozoa Fungi

Code view:

```
for each abc in db
if abc="Algae" then
%>
      <a href="#algae">Algae</a>

<% elseif abc="Bacteria" then %>
      <a href="#bacteria">Bacteria</a>

<% elseif abc="Virus" then %>
      <a href="#virus">Virus</a>

<% elseif abc="Protozoa" then %>
      <a href="#protozoa">Protozoa</a>

<% elseif abc="Fungi" then %>
      <a href="#fungi">Fungi</a>

<% elseif abc="HostParasite" then %>
      <a href="#hostparasite">HostParasite</a>

<% elseif abc="Zingiberaceae" then %>
      <a href="#ginger">Zingiberaceae</a>

<% elseif abc="Fern" then %>
      <a href="#fern">Fern</a>

<% elseif abc="Bactiden" then %>
      <a href="#bactiden">Bactiden</a>

<% elseif abc="Photogallery" then %>
      <a href="#photogallery">Photogallery</a>

<% elseif abc="sef_flies" then %>
      <a href="#sef_flies">Fruit Flies</a>
```

Fig. 6.19 Interface view and codes for linking results from other databases that were selected. This feature expedites the navigation in results page

images were shrunk to a smaller size for faster download and ease of view. The shrinking of images is done by setting a standard size in the HTML codes to view the results. However these images can be easily zoomed out for a clearer view by clicking them. Results were obtained in 0.09 s and 0.11 s respectively.

6.3.5 Flexible Data Format

The data format used in CABIR is generic to all fields in all the databases tested (see Table 5.1). However, it can be altered according to the fields in the databases. The data format is transformed into XML to act as a wrapper in the CABIR architecture (presented in Chapter 7). The fields in this XML wrapper can be added and removed according to the contents of the databases shared.

Interface view:

Family

1	**Family:** Chlorophyta **Genus:** Chlorophyceae (Chlorococcales) **Species:** Chlorella vulgaris Beijerinck
2	**Family:** Clorophyta **Genus:** Chlorophyceae (Chlorococcales) **Species:** Chlorococcum sp.
3	**Family:** Cyanophyta **Species:** Chroococcus sp
4	**Family:** Chlorophyta **Genus:** Chlorophyceae (Trentepohliales) **Species:** Trentepohlia sp.

Code view:

```
<% count=count + 1
i_last=right(count,1)
select case i_last
case "1","3","5","7","9"
bgcolor = "#C0E4C6"
case else
bgcolor = "#FFFFFF"
end select
%>
<table bgcolor=<%=bgcolor%> border="0" cellpadding="0" cellspacing="0" style="border-collapse: collapse" bordercolor="#111111" width="100%">
<tr valign=top>
<td rowspan=67>
<b>
<font color=blue>
<%=Num%>
</font>
</b> </td> </tr>
```

Fig. 6.20 Interface view and codes separates the results of query. When case count = odd number, table is in *green* whereas when case = even number, table is in *white color*. This feature makes view of results clearer

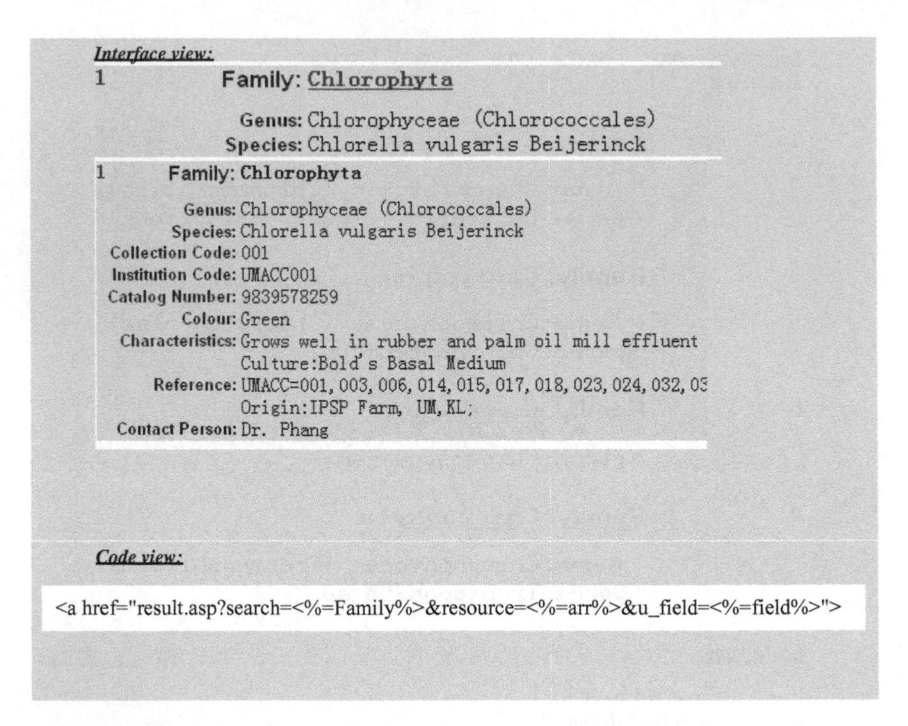

Fig. 6.21 Details of a record are embedded in the hyperlinked Family. Users can view details of a particular record by clicking on the hyperlink. The code view show the ASP codes implemented for this function

6.3.6 Link to Remote Cross Platform Providers

CABIR can link databases in Windows and UNIX based operating systems. The relational databases tested with CABIR include both from Windows and UNIX platforms. The Fruit Flies database resides in Solaris 9 operating system. The Biodiversity sample database resides in Linux operating system at MIMOS, Technology Park Malaysia. The search results from these databases are presented in Figs. 6.20 and 6.21.

6.3.7 Link to Heterogeneous Database Management Systems

Linking heterogeneous databases is one of the important features of CABIR. Database Management Systems tested with CABIR in this research are Microsoft Access, MySQL, FileMaker and DB2. These are some of the most commonly used DBMSs (see Chapter 2). Figures 6.25 and 6.26 display results from mySQL databases. Figures 6.22, 6.23 and 6.24 display results from FileMaker DBMSs. Figures 6.27, 6.28, 6.29 displays results from DB2 DBMS while Fig. 6.34 displays

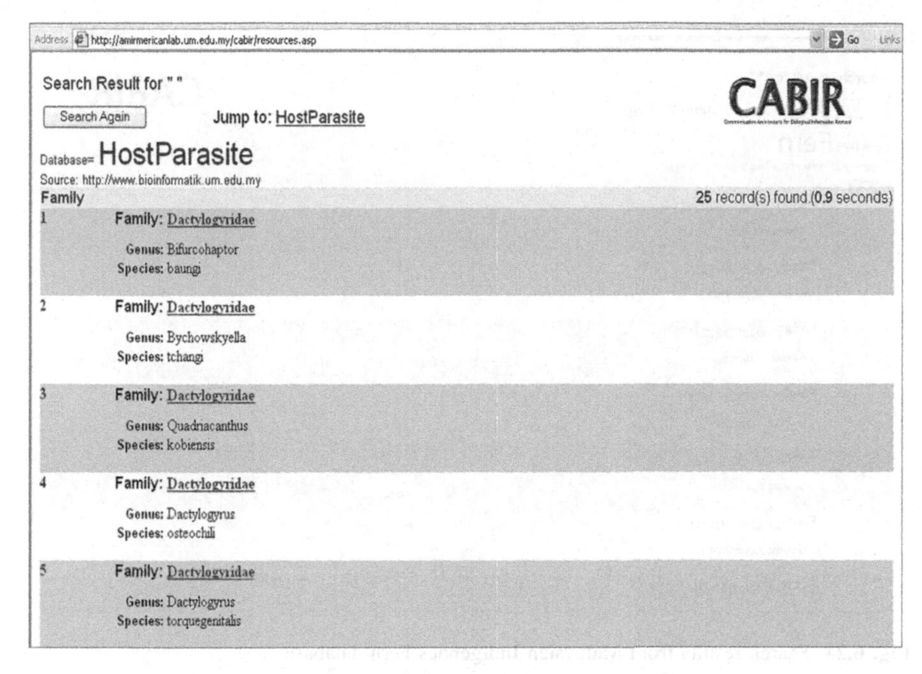

Fig. 6.22 Search results from Host Parasite database

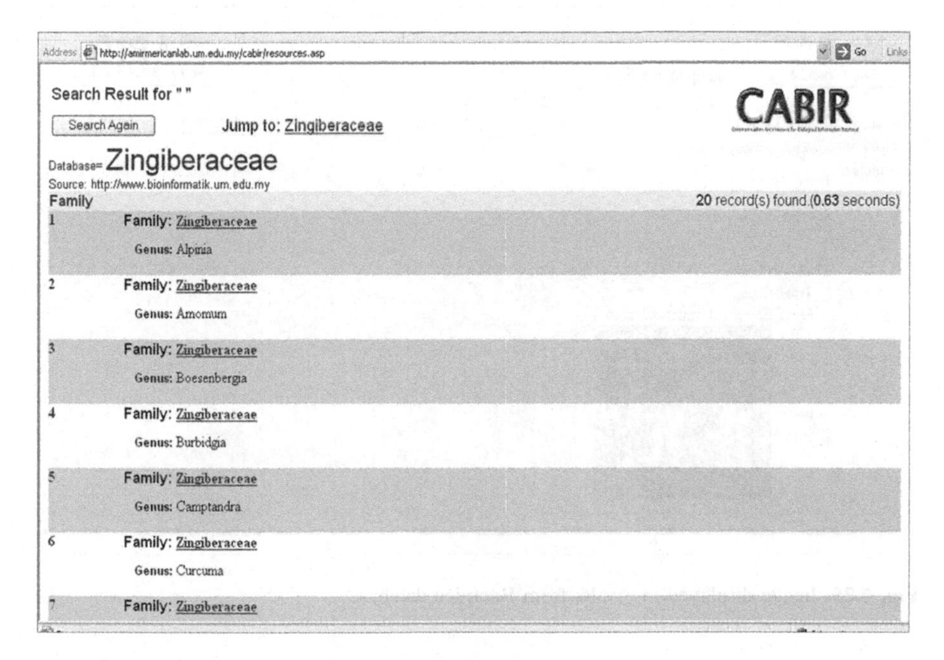

Fig. 6.23 Search results from Zingiberaceae database

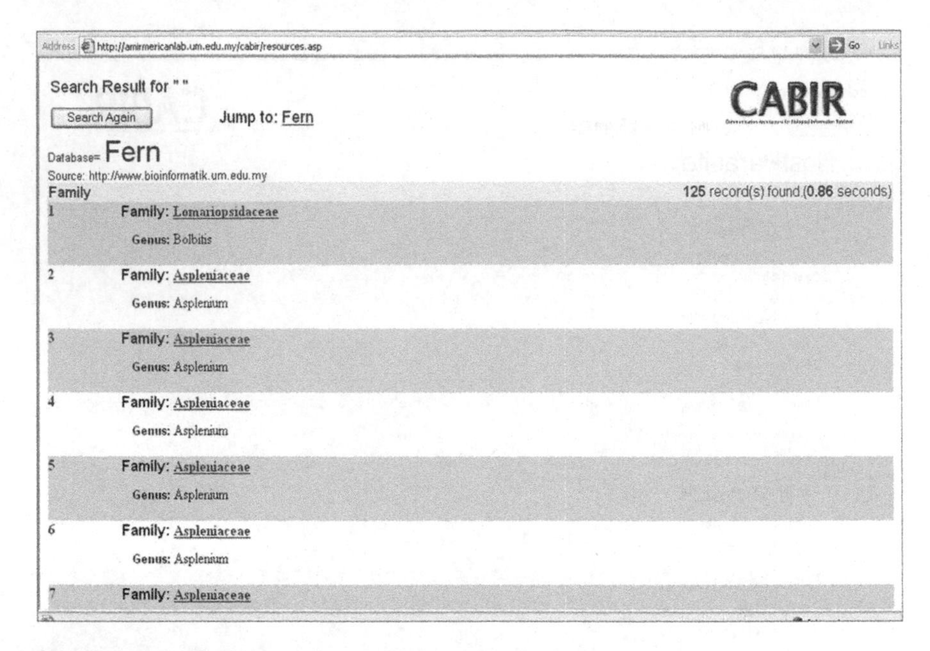

Fig. 6.24 Search results from Malaysian Indigenous Fern database

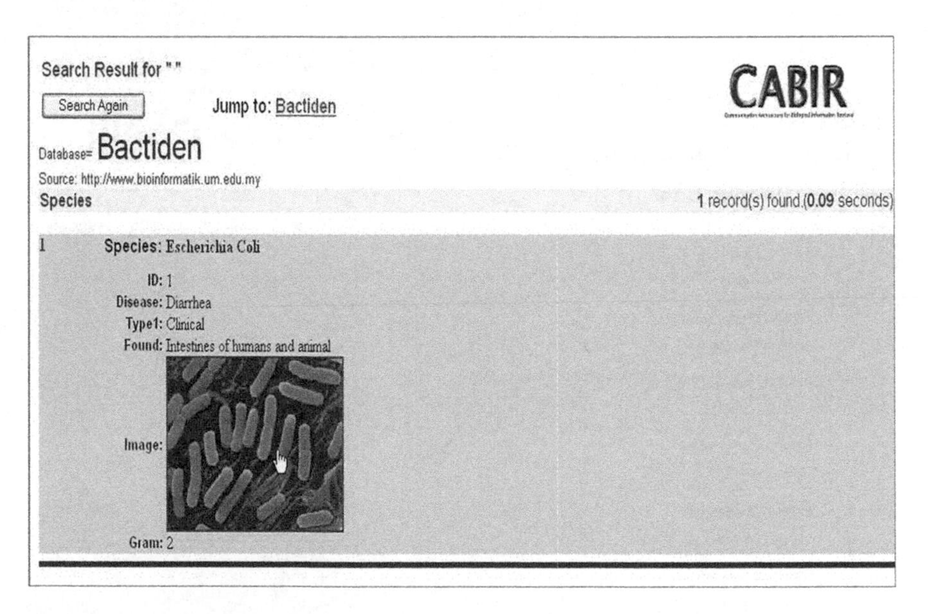

Fig. 6.25 Image displayed in results from Bactiden database

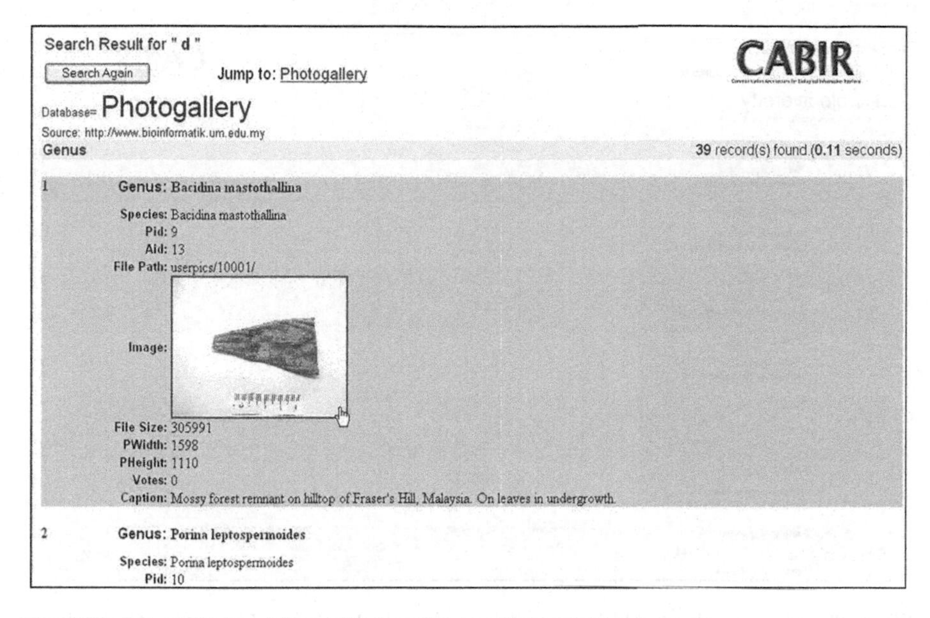

Fig. 6.26 Image displayed in results from Photogallery database

1
Species: Bactrocera longivittata
Geographical Distribution: Selangor
Subfamily: Dacinae
Tribe: Dacini

2
Species: Eupharanta figurata
Geographical Distribution: Sarawak
Subfamily: Dacinae
Tribe: Dacini

3
Species: Carpophthorella soolcao
Geographical Distribution: Peninsular Malaysia
Subfamily: Dacinae
Tribe: Dacini

4
Species: Gastrozona selangorensis
Geographical Distribution: Peninsular Malaysia
Subfamily: Dacinae
Tribe: Dacini

5
Species: bactrocera malaysiensis
Geographical Distribution: Peninsular Malaysia, Singapore
Subfamily: Dacinae
Tribe: Dacini

6
Species: Bactrocera calambolae

Fig. 6.27 Results of all records from fruit flies database in Solaris 9 operating system

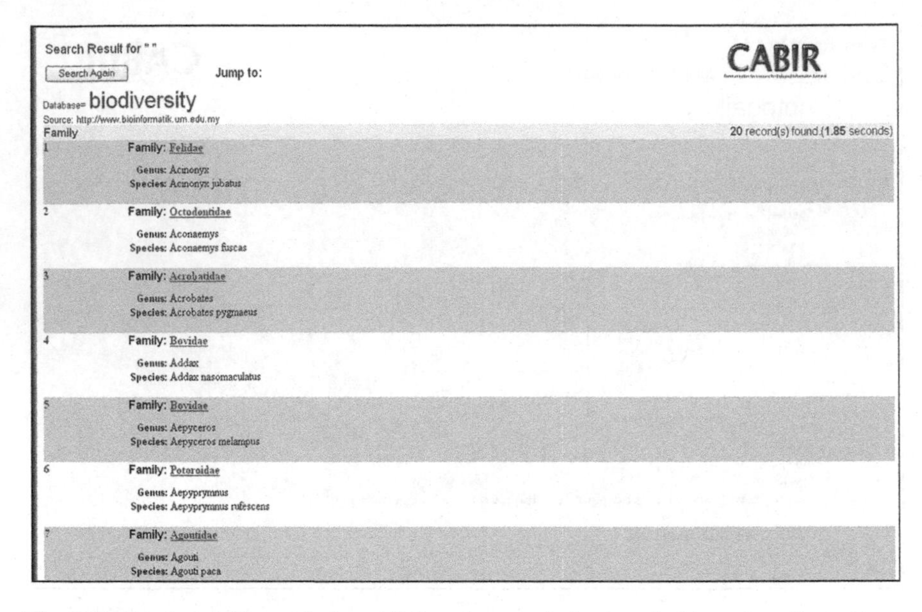

Fig. 6.28 Results of all records from biodiversity sample database in Linux operating system

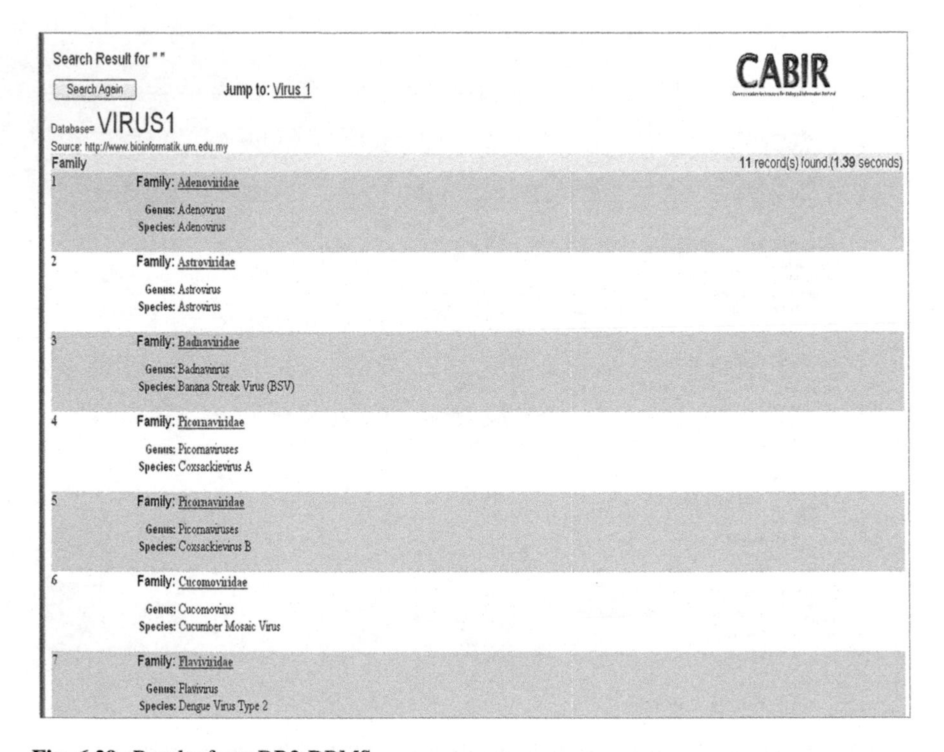

Fig. 6.29 Results from DB2 DBMS

results from Microsoft Access DBMS. Results from DB2 DBMS were obtained in 1.39 s while results from Microsoft Access database were obtained in 0.09 s.

6.3.8 *Error Messages*

CABIR gives the user an insight to errors by giving informative messages. For example, if a user typed a text in the search textbox which does not exist in the selected database(s), there will be a “No records found in this database.” message (see Fig. 6.30).

Some databases provide only a limited set of searchable fields. The Zingiberaceae database allows searches only on Family and Genus, the Bactiden database on Species, and the Photogallery on Genus and Species. If a user searches for a field that does not exist in a database, there will be a “There was an error.” message (see Fig. 6.31).

6.4 System Design

In this section, CABIR is discussed as an overall system for database integration. The subsections below show and discuss the main screens in CABIR.

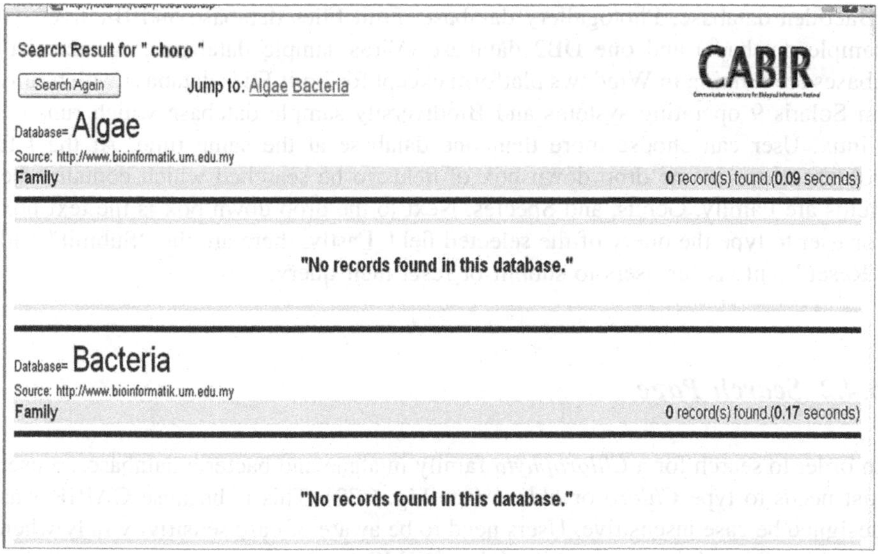

Fig. 6.30 Error message (“No records found”)

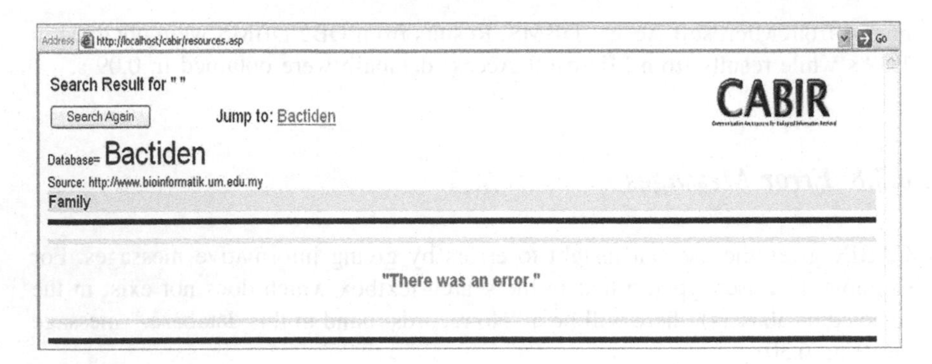

Fig. 6.31 Error message ("There was an error")

6.4.1 Main Page

Figure 6.32 shows the main page of the CABIR system. There is a logo of CABIR on the center of the page. The left and right logos represent University Malaya and Institute of Biological Sciences respectively. Underneath is the form for collecting user query. There are thirteen databases at the right column of the form, from which a user can choose to send query. Among them, five are Microsoft Access databases (Algae database, Bacteria database, Virus database, Protozoa database, and Fungi database), three are FileMaker databases (Host Parasite database, Zingiberaceae database, and Fern database), four more are MySQL databases (Bactiden database, Photogallery database, Fruit Flies database and Biodiversity sample database) and one DB2 database (Virus sample database). All the databases are running in Windows platform except for Fruit Flies database which runs on Solaris 9 operating systems and Biodiversity sample database which runs on Linux. User can choose more than one database at the same time. At the left column, there is one drop down box of fields to be searched which contains the fields are Family, Genus, and Species. Next to the drop down box is the text box for user to type the query of the selected field. Lastly, there are the "Submit" and "Reset" buttons for users to submit or reset their query.

6.4.2 Search Page

In order to search for a *Chlorophyta* family in algae and bacteria databases, a user just needs to type *Chloro* or *chloro* (see Fig. 6.32). This is because CABIR was designed be case insensitive. Users need to be aware of case sensitivity only when searching from databases on FileMaker DBMSs.

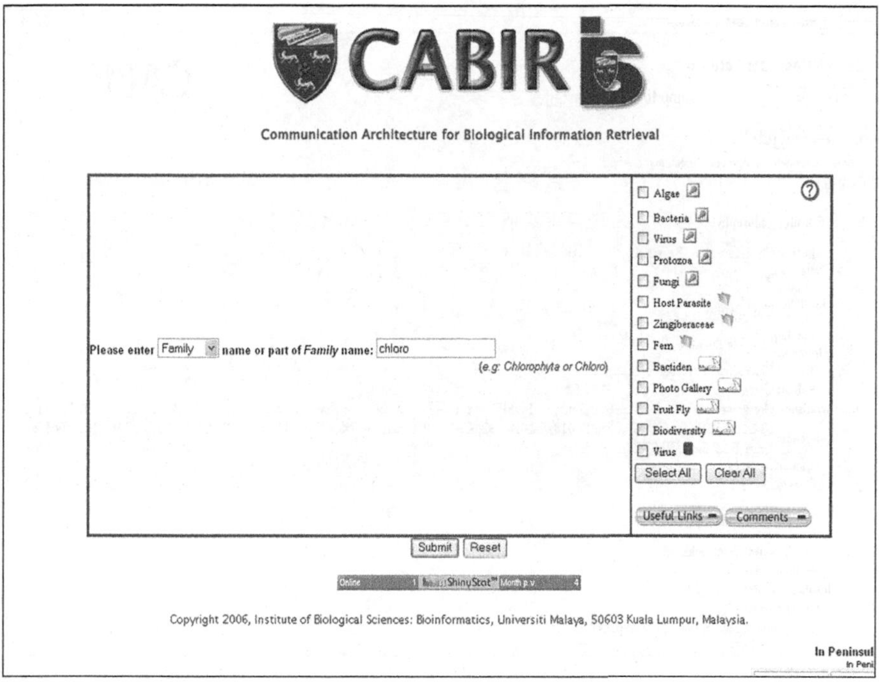

Fig. 6.32 Main page of CABIR

6.4.3 Results of Search

An example of search in CABIR is displayed in Fig. 6.33. There is a "Search Again" button at the top of the page, which allows the user to go back to the search page. Next to the button is the "Jump to" navigational links, which enable user to navigate to the next databases to look for the search results. There is one header for every database which describes the source of database (location of database). There is a bar below the header showing the field user is searching from and number of records found. Each record is segregated by the alternate green and white background. The name of the selected field is made bold so that it is easily seen by user. Results were obtained in 0.09 s.

6.4.4 Searching all Records Within a Database

In order to search for all the records in a database, a user just have to leave the text box empty and press the "Submit" button. (see Fig. 6.34).

(a) Example of search all records from existing databases.

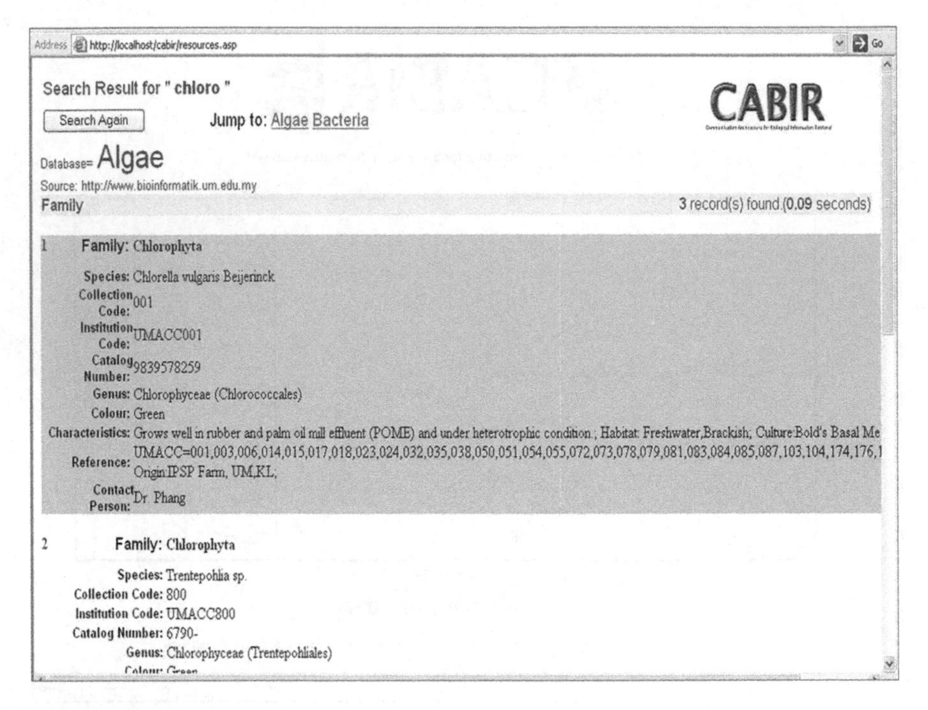

Fig. 6.33 Results of search

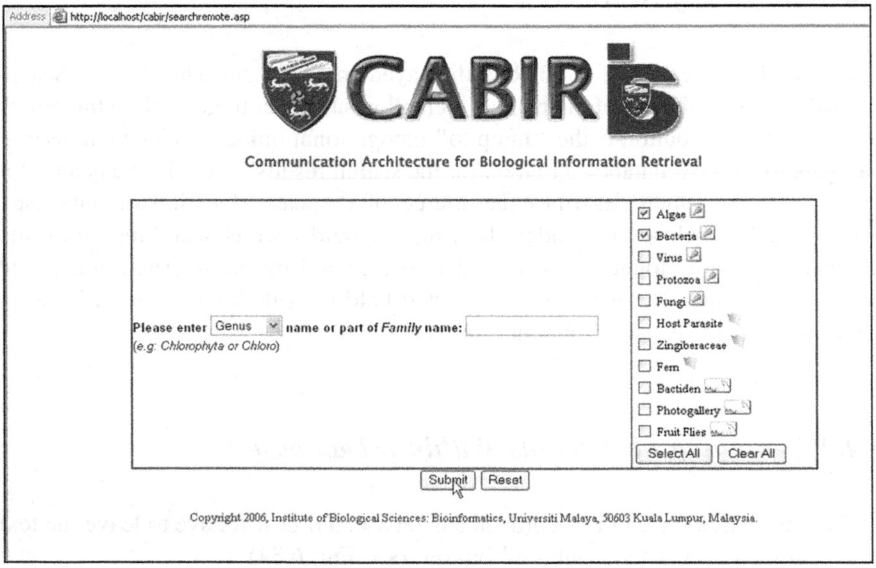

Address http://localhost/cabir/searchremote.asp

CABIR

Communication Architecture for Biological Information Retrieval

Please enter Genus name or part of *Family* name:
(*e.g: Chlorophyta or Chloro*)

☑ Algae
☑ Bacteria
☐ Virus
☐ Protozoa
☐ Fungi
☐ Host Parasite
☐ Zingiberaceae
☐ Fern
☐ Bactiden
☐ Photogallery
☐ Fruit Flies

Select All Clear All

Submit Reset

Copyright 2006, Institute of Biological Sciences: Bioinformatics, Universiti Malaya, 50603 Kuala Lumpur, Malaysia.

Fig. 6.34 Searching all records in Algae and Bacteria databases (search box is left empty)

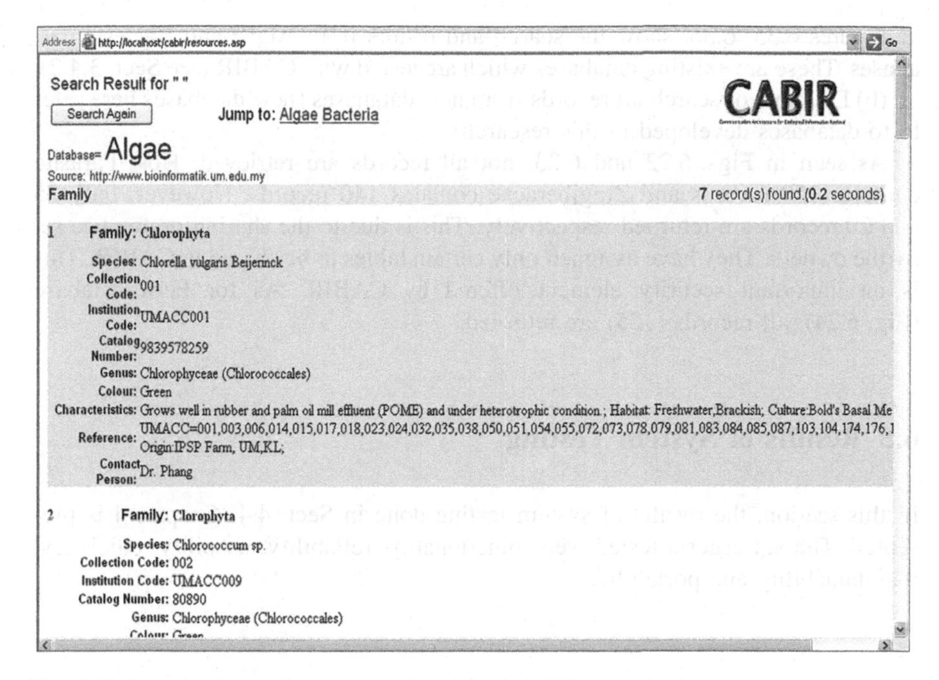

Fig. 6.35 Results of searching all records in Algae and Bacteria databases

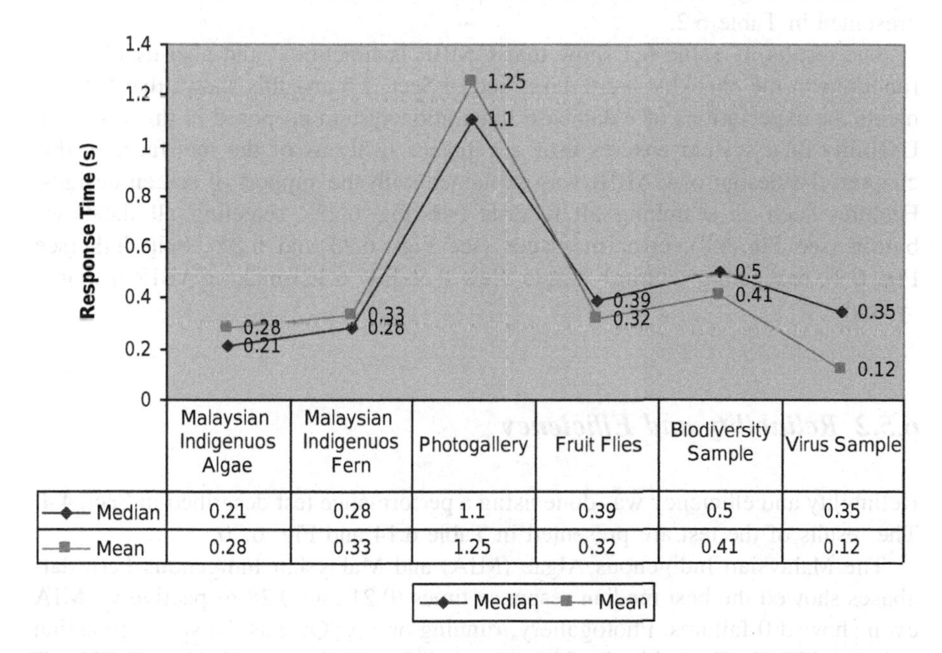

	Malaysian Indigenuos Algae	Malaysian Indigenuos Fern	Photogallery	Fruit Flies	Biodiversity Sample	Virus Sample
Median	0.21	0.28	1.1	0.39	0.5	0.35
Mean	0.28	0.33	1.25	0.32	0.41	0.12

Fig. 6.36 Response time mean and median plots for six databases using the CABIR system

Figures 6.35, 6.36, show the search and results from Algae and Bacteria databases. These are existing databases which are tested with CABIR (see Sect. 3.4.2).

(b) Example of search all records from new databases (new databases here refer to to databases developed in this research).

As seen in Figs. 6.22 and 6.23, not all records are retrieved. Host Parasite contains 150 records and Zingiberacae contains 140 records. However, only 25 and 20 records are returned respectively. This is due to the sharing preference set by the owners. They have assigned only certain tables to be shared in CABIR. This is an important security element offered by CABIR. As for Fern database (Fig. 6.24), all records (125) are returned.

6.5 Results of System Testing

In this section, the results of system testing done in Sect. 4.4, Chapter 4 is presented. The six criteria tested were functionality, reliability, usability, efficiency, maintainability and portability.

6.5.1 Functionality and Usability

The functionality and usability test was performed with a checklist and results are presented in Table 6.2.

The results in Table 6.2 show that CABIR is functional and also usable. The functions in the checklist were described in Sect. 1.6 and this shows that CABIR meets the expectations of a database integration system proposed in this research. Usability of a system ensures user acceptance in terms of the interface. In this chapter, the design of CABIR was explained with the support of screen designs. Features such as searching all records (see Fig. 6.28), selecting all databases button (see Fig. 6.8), error messages (see Figs. 6.23 and 6.24), help link (see Fig. 6.9), navigating within a single view (see Fig. 6.12) make CABIR a usable system.

6.5.2 Reliability and Efficiency

Reliability and efficiency was done using a performance test described in Sect. 4.4. The results of the test are presented in Table 6.14 and Fig. 6.30.

The Malaysian Indigenous Algae (MIA) and Malaysian Indigenous Fern databases showed the best median response times (0.21 and 0.28 respectively). MIA even showed 0 failures. Photogallery, running on mySQL has the slowest median response time (1.1), with the highest number of failures (3). This is because

Table 6.2 Functionality and usability checklist results

Criteria	Yes	No
Functionality		
Link remote databases in a networked environment	/	
Support heterogeneous data format	/	
Support heterogeneous database management systems	/	
Link databases hosted in Windows and UNIX based platforms	/	
Provide data security for database owners by allowing them to keep and maintain their own data and to choose information to be shared and linked	/	
Usability		
It is easy to know the current location within the overall map of the site	/	
It is clear what information is available at the current location	/	
It is clear where you can go from the current location	/	
It is always easy to return to the Home Page	/	
Links are used and appear in standard web style	/	
Menus are used and appear in standard web style	/	
Link labels match destination page titles or headers	/	
Labels and links are described clearly	/	
Color choices allow for easy readability	/	
If necessary, error messages are clear and in plain language	/	
The content language is clear and simple	/	

Photogallery is an image database and due to the size of images, it takes longer to retrieve them. However, it is difficult to generalize about these results as the performance of a data source actually depends on number of factors, such as the server hardware and software, the database design and the network performance. For the six databases queried, the operating systems used include both UNIX and Windows based and the Web servers were Apache and IIS. Nevertheless, it is indeed very promising that five out of six databases have median response time of less than a second. This means that the response time for search using the CABIR system is less than a second and therefore CABIR is efficient. In terms of reliability, the highest number of failures obtained is 3 and this is very low compared to the number of times it was searched, which are 100. This concludes that CABIR is reliable as well (Table 6.3).

6.5.3 Maintainability and Portability

The maintainability and portability of CABIR was determined by the underlying architecture of CABIR. Stability is a sub-characteristic of maintainability (see Table 3.5), therefore in this section, maintainability in terms of stability is discussed.

CABIR has two main parts, the portal and providers. There is a single portal communicating with distributed and heterogeneous providers (see Fig. 6.6). The

Table 6.3 Performance test results for six databases. Each database was queried 100 times the species names. The table displays the median, mean and best and worst times taken for a database to respond to a query. The number of times a query failed to return a response is also recorded

Source	Response time (in seconds)				
	Median	Mean	Best	Worst	Failed
Malaysian indigenous algae	0.21	0.28	0.18	1.88	0
Malaysian indigenous fern	0.28	0.33	0.55	1.21	2
Photogallery	1.1	1.25	0.84	1.28	3
South East Asia fruit flies	0.39	0.32	0.22	1.62	2
Biodiversity sample database	0.5	0.41	0.95	1.9	5
Virus sample database	0.85	0.12	0.05	1.52	1

portal was developed using ASP whereas the providers in ASP or PHP. The programming tools and techniques make CABIR a very stable system.

ASP is the main language used in this research due to some important factors:-

1. The data integration system in this research was developed using Microsoft Windows 2003 server and ASP is a Microsoft's language. It is therefore the best combination.
2. Most database integration systems such as DiGIR are developed on open source platform using PHP. There is a need to develop a Windows based database integration system using ASP.
3. ASP is compatible with a lot of scripting languages. For example, a PHP page can be invoked in ASP.

XML and ASP are two main languages used in this research to build CABIR. It is a very efficient and a powerful combination. XML is a simple and a powerful tool for Web developers. XML was created to handle complex Web documents. XML allows programmer to define all his own tags with rules such as data description and data relationships. XML is used in order to remove the cumbersome problems that were faced with HTML. Information can be accessed easier using XML. ASP and XML are powerful tools for creating dynamic Web pages (see http://www.xml-training-guide.com/asp-xml.html) (see Chapter 3 for literature review on ASP, XML and HTML).

ASP uses XML as a tool in its application for a simple reason that data in HTML is allowed to transmit between dissimilar platforms (portable), whereas XML allows one to express complex structure. Moreover it allows the programmer to create his own tags with all sorts of rules. A new document instance can be created using MSXML. There are a number of ways to access XML data from an ASP page. Document Object Model (DOM) plays an important role in retrieving the XML data from ASP. In DOM a document is viewed as a tree of nodes. Every node of the tree can be accessed randomly. The main advantage is that it provides all functions in an object based way. An XML parser based on DOM from Microsoft is MSXML. This component is used for accessing the XML documents.

These tools made CABIR a portable and maintainable application for database integration.

Another sub-characteristic of maintainability is changeability (see Table 4.6). CABIR can be easily changed by adding providers and wrappers into the system. Each data source (relational database) in CABIR has a provider and an xml wrapper. They are independent entities within the loosely coupled architecture. Therefore, CABIR can handle multiple distributed relational and heterogonous databases.

6.6 Conclusion

In this chapter, the features of CABIR were presented with the support of screen designs of the system. From the results, it can be concluded that this system has met the requirements of the proposed database integration system in Sect. 1.6. The features of CABIR are summarized and discussed against DiGIR in Chapter 7.

References

Vieglais, D. (2003). The Darwin Core (online) SpeciesAnalyst. Available from: http://speciesanalyst.net/docs/dwc/index.html. Accessed 12 January 2005.

Sarinder, KKS., Majid, MA., Lim, LHS., Ibrahim, H & Merican, AF. (2006). Living Web (online). University Malaya, Kuala Lumpur. Available from: http://umbioWeb.um.edu.my. Accessed 1 July 2006.

Sarinder, KKS., Majid, MA., Lim, LHS., Ibrahim, H & Merican, AF. (2005). integrated biological database initiative (IBDI). In: Proceedings of International Conference on Biogeography and Biodiversity Wallace in Sarawak—150 years later, Kuching, Malaysia.

West, L. (2001). Net HTTP client (online). Available from: http://lwest.free.fr/doc/php/lib/net_http_client-en.html. Accessed 5 December 2005.

Chapter 7
Concluding Remarks

7.1 Introduction

Interoperability among heterogeneous databases has long been seen as being of major importance (Litwin et al. 1990; Sheth and Larson 1990). Various kinds of federated architectures are employed, depending on the nature of the databases involved. In particular wrapping techniques are often used to remove some elements of heterogeneity (Roth and Schwarz 1997) and mediators are used to draw together information from disparate sources (Wiederhold 1992). A database that does not follow a standard data model requires extensive use of metadata for mapping and a study by Baru et al. (1999) showed that an increasingly common way of providing this is by use of XML. In this research, mediator based database integration system with XML wrappers was developed.

Referring back to Chapter 1, the aim of this research was to develop an indigenous framework for integrating distributed heterogeneous and relational biodiversity databases to facilitate biodiversity data sharing among the scientific community especially Malaysia as well as the rest of the world. Data security and simplicity were few of the main concerns of this system. With the evaluation done in Chapter 4, it was noted that CABIR fulfills the aim of this research.

7.2 CABIR in Biodiversity Databases

This research produced results showing that CABIR is able to integrate the different views and versions of taxonomic data, making it available in simple formats, with friendly interfaces to be shared among the scientific community and to bring them together to work as a team to sustain global biodiversity heritage, without expecting them to export their collected information to a centralized location (see Chapter 1).

S. K. Dhillon and A. S. Sidhu, *Data Intensive Computing for Biodiversity*,
Data, Semantics and Cloud Computing 485, DOI: 10.1007/978-3-642-38047-1_7,

7.3 CABIR as a Simple and Dynamic Database Integration System

Due to the powerful programming tool and techniques, the underlying details of CABIR are concealed from users. Users do not need to know where the databases reside, the structure of the system, Database Management System used to build databases and the programming behind the system. The information retrieval is similar to a system accessing a local database. This is to ensure its simplicity.

CABIR's user interface is made simple and straightforward. A user does not need much effort in understanding the functionality of the system. Thus it will suit a wide spectrum of users. Every page is designed to be as simple and as compact as possible. These pages load in a fast response time to ensure that users do not need wait long to view the pages.

CABIR system is designed to support search in multiple heterogeneous databases simultaneously. Heterogeneous here means the databases are developed using a variety of Database Management Systems and in heterogeneous data format (see Sects. 6.3.5 and 6.3.7). CABIR also works for providers in UNIX based systems. It has been tested with a MySQL database in Solaris 9 platform and it has shown success (see Sect. 6.3.6).

7.4 CABIR and Data Security

As providers of CABIR, databases owned by scientists are secured. A database owner can choose the data to be shared with users in the public domain. This can be achieved by configuring the provider file. As seen in Sect. 6.4.4, not all records must be returned in the query. Furthermore, database owners do not need to export their data into a central warehouse in order to share their data. Instead they can store their data in their own hosts, while allowing them to be shared. Thus, they are able to protect their databases. Basically, CABIR brings forth interested users to share data while protecting confidential information.

7.5 A Summary of Comparison Between CABIR and DiGIR

CABIR is a database integration system based on the generic characteristics of existing systems (see Table 3.1). Out of these systems, DiGIR was implemented in the preliminary study in Chapter 2. However, due to some predicaments, a new database integration system, CABIR was developed in this research (see Chapter 2 for more details).

CABIR has some differences compared to DiGIR. CABIR is a Microsoft based system developed using IIS Web server, ASP, XML, JavaScript, VBScript and SQL whereas DiGIR is an open source system which uses the Apache Web server with PHP and XML. CABIR is specially designed for both low level uses as well as high level uses as it stresses simplicity and usability. It adopts the ODBC, DataDirect32-BIT SequeLink 5.4 and Oledb database connectivity. On the contrary, DiGIR only uses Adodb database connectivity. This limits it from reading FileMaker databases. Another difference is that DiGIR is built to accommodate the Darwin Core schema or data standard which is slightly rigid when mapping to indigenous databases. Databases were altered when implementing the DiGIR system as they did not have the required fields such as *Date Last Modified, Institution Code, Collection Code, Catalogue Number and Collection Code* (see Chapter 2). These fields are required by Darwin Core data format. CABIR on the other hand does not restrict to Darwin Core data format. It uses a format which can be extended when needed. Typically it is an extension of the Darwin Core v2 data format.

In CABIR, database owners can choose with whom they would like to share their data with but in DiGIR, once a database is submitted as a DiGIR provider, the database will automatically be shared with all other DiGIR providers in the world.

7.6 CABIR as a Model Database Integration System

CABIR works well on biodiversity databases and it may also work for data in other biological domains and for a variety of different data sets outside the biological domain. This is because CABIR's components are independent of each other, especially the data format. Therefore, different data formats can be used with CABIR.

Besides that, CABIR also has generic characteristics of existing database integration systems (see Table 7.1). These systems were used as models to build CABIR, especially DiGIR which was implemented during the preliminary study. Generally, CABIR satisfies the need of a generic database integration system. In addition, it is made simple with powerful underlying facilities. Thus, it is suitable for scientific community. In summary, CABIR can be used as one of the standard approach towards database integration.

7.7 CABIR Limitations

Despite the above strengths, CABIR is has its own weaknesses. Several weaknesses are: (1) CABIR rely on network for data access. Network problems include bottlenecks, low response times and the occasional unavailability of sources (2) Search fields currently available in the system are restricted to Family, Genus,

Table 7.1 CABIR shows match in criteria of a standard database integration system (see Sect. 2.2)

	Aim of integration	Data model	Source model	User model	Level of transparency	Overall integration approach
GenoMax	Data mining	Structured static data	Mostly complementary	Expertise in software functionality, data mining tools, life science informatics analysis approaches, collaboration work, and other aspects of the user interface	Sources specified by head database	Warehouse based
Kleisli	Query-oriented	Semi-structured, object-oriented	Mostly complementary	Expertise in query language	Sources specified by user	Mediator-based
DiscoveryLink	Query-oriented middleware	Structured, object-relational	Mostly complementary, some overlap	Expertise in query language	Sources selected by system	Mediator-based
SPICE	Query-oriented middleware	Structured, object-relational	Mostly complementary	Expertise in query language	Sources specified by user	Mediator-based
DiGIR	Query-oriented	Structured, object-relational	Mostly complementary	Expertise in query language	Sources specified by user	Mediator-based
CABIR	Query-oriented	Structured, object-relational	Mostly complimentary can also accommodate other kind of data with a change of data model	Novice	Sources specified by user	Mediator-based

and Species (3) The system is unable to inform users when users input wrong characters, it only show "No records found in this database" (4) Insertion of providers must be done manually (5) Although the providers in CABIR hosted either in Windows or Unix, the portal is strictly developed for Windows servers.

7.8 Future Focus

The following can be implemented as future research:

1. Testing CABIR using other biological and non-biological databases.
2. Development of an application with friendly user interface to add providers in CABIR. The scope of this research (see Chapter 1) did not involve automated software for database owners to add their databases (providers) automatically into the system.
3. Linking CABIR with DiGIR, since DiGIR is widely used by GBIF, CABIR can be integrated with DiGIR to allow wider spectrum of users.
4. Development of a UNIX version of CABIR portal.
5. Adding more search fields.
6. Improved feedback for wrong entries in the search box.

7.9 Conclusion

CABIR is a simple system designed to serve the purpose of storing, disseminating and sharing biodiversity information among the scientific community. It also provides interaction among researchers while maintaining the privacy of their databases by controlling the permission for sharing. They are also able to host and maintain their own databases. CABIR is able to integrate databases in various DBMSs including FileMaker. CABIR is also able to retrieve images from the integrated databases and search results are typically returned in few seconds in CABIR. Therefore, it is a high performance database integration system while maintaining its simplicity and having the generic characteristics of existing database integrations. This system can be further explored and improved to handle the up to date demands of the scientific community. CABIR is expected to work with data sets outside the biological domain due to its flexibility. Finally, CABIR is deemed to attract various researchers interested in database integration field.

References

Baru, C., Chu, V., Gupta, A., Ludäscher, B., Marciano, R., Papakonstantinou, Y., Velikhov, P. (1999). XML-based information mediation for digital libraries. In E. A. Fox & N. Rowe (Eds.), *Proceedings of the Fourth ACM Conference on Digital Libraries* (pp. 214–215). New York: ACM Press.

Litwin, W., Mark, L., & Roussopoulos, N. (1990). Interoperability of multiple autonomous databases. *ACM Computing Surveys (CSUR),* 22(3) , pp 267-293.

Roth, M. T., & Schwarz, P. M. (1997). Don't scrap it, warp it! A wrapper architecture for legacy data sources. In *VLDB'97, Proceedings of 23rd International Conference on Very Large Data Bases*, pp. 266–275.

Sheth, A. P., & Larson, J. A. (1990). Federated database systems for managing distributed, heterogeneous, and autonomous databases. *ACM Computing Surveys, 22*(3), 183–236.

Wiederhold, G. (1992). Mediators in the architecture of future information systems. *IEEE Computer, 25*(3), 38–49.

Appendix A
Survey for Developing Standard for Biodiversity Databases

Name :
Email Address :
Research Area :
Dear Sir/Madam, I am conducting a survey on data formats for developing biodiversity databases (plant, animal, microorganisms etc). Due to this, I would like to have your opinion on the attributes needed in the databases. Please tick the boxes below for fields which are needed in the databases. Your time and effort will be very much appreciated.

Name	**Name**
Date last modified	Serotype
Institution code	State province
Collection code	Locality
Catalog number	Longitude
Scientific name	Latitude
Basis of record	Coordinate precision
Kingdom	Bounding box
Phylum	Minimum elevation
Class	Maximum elevation
Order	Minimum depth
Family	Maximum depth
Genus	Sex
Species	Preparation type
Subspecies	Individual count
Scientific name author	Previous catalog number
Identified by	Relationship type
Year identified	Related catalog item
Month identified	Notes
Day identified	OTHER SUGGESTIONS/RECOMMENDATIONS
Type status	
Collector number	
Field number	
Collector	
Year collected	
Month collected	
Day collected	
Continent ocean	
Country	

S. K. Dhillon and A. S. Sidhu, *Data Intensive Computing for Biodiversity*, Data, Semantics and Cloud Computing 485, DOI: 10.1007/978-3-642-38047-1,

Appendix B
Darwin Core V2 Searchable Concepts

Concept ID	Name	Required	Type	Description
xx	Date last modified	Y	Date time	ISO 8601 compliant stamp indicating the date and time in UTC(GMT) when the record was last modified. Example: the instant "November 5, 1994, 8:15:30 am, US Eastern Standard Time" would be represented as "1994-11-05T13:15:30Z" (see W3C Note on Date and Time Formats). (What to do when this date-time is unknown? Use Date-Time first "published"?
10	Institution code	Y	Text	A "standard" code identifier that identifies the institution to which the collection belongs. No global registry exists for assigning institutional codes. Use the code that is "standard" in your discipline
11	Collection code	Y	Text	A unique alphanumeric value which identifies the collection within the institution
12	Catalog number	Y	Text /numeric	A unique alphanumeric value which identifies an individual record within the collection. It is recommended that this value provides a key by which the actual specimen can be identified. If the specimen has several items such as various types of preparation, this value should identify the individual component of the specimen

(continued)

S. K. Dhillon and A. S. Sidhu, *Data Intensive Computing for Biodiversity*, Data, Semantics and Cloud Computing 485, DOI: 10.1007/978-3-642-38047-1,

(continued)

Concept ID	Name	Required	Type	Description
1	Scientific name	Y	Text	The full name of lowest level taxon the Cataloged Item can be identified as a member of; includes genus name, specific epithet, and subspecific epithet (zool.) or infraspecific rank abbreviation, and infraspecific epithet (bot.) Use name of suprageneric taxon (e.g., family name) if Cataloged Item cannot be identified to genus, species, or infraspecific taxon
xx	Basis of record	N	Text	An abbreviation indicating whether the record represents an observation (O), living organism (L), specimen (S), germplasm/seed (G), etc
2	Kingdom	N	Text	The kingdom to which the organism belongs
3	Phylum	N	Text	The phylum (or division) to which the organism belongs
4	Class	N	Text	The class name of the organism
5	Order	N	Text	The order name of the organism
6	Family	N	Text	The family name of the organism
7	Genus	N	Text	The genus name of the organism
8	Species	N	Text	The specific epithet of the organism
9	Subspecies	N	Text	The sub-specific epithet of the organism
xx	Scientific name author	N	Text	The author of a scientific name. Author string as applied to the accepted name. Can be more than one author (concatenated string). Should be formatted according to the conventions of the applicable taxonomic discipline
xx	Identified by	N	Text	The name(s) of the person(s) who applied the currently accepted Scientific Name to the Cataloged Item
xx	Year identified	N	Numeric	The year portion of the date when the Collection Item was identified; as four digits (−9999..9999), e.g., 1906, 2002
xx	Month identified	N	Numeric	The month portion of the date when the Collection Item was identified; as two digits (01..12)
xx	Day identified	N	Numeric	The day portion of the date when the Collection Item was identified; as two digits (01..31)

(continued)

(continued)

Concept ID	Name	Required	Type	Description
xx	Type status	N	Text	Indicates the kind of nomenclatural type that a specimen represents. (This is incomplete because type status actually describes the relationship between a name and a specimen (or ternary relatiohnship between a specimen, name, and publication).) In particular, the type status may not apply to the name listed in the scientific name, i.e., current identification. In rare cases, a single specimen may be the type of more than one name
xx	Collector number	N	Text	An identifying "number" (really a string) applied to specimens (in some disciplines) at the time of collection. Establishes a links different parts/preparations of a single specimen and between field notes and the specimen
xx	Field number	N	Text	A "number" (really a string) created at collection time to identify all material that resulted from a collecting event
13	Collector	N	Text	The name(s) of the collector(s) responsible for collection the specimen or taking the observation
14	Year collected	N	Numeric	The year (expressed as an integer) in which the specimen was collected. The full year should be expressed (e.g. 1972 must be expressed as "1972" not "72")
15	Month collected	N	Numeric	The month of year the specimen was collected from the field. Possible values range from 01…12 inclusive
6	Day collected	N	Numeric	The day of the month the specimen was collected from the field. Possible value ranges from 01..31 inclusive
4	Julian day	N	Numeric	The ordinal day of the year; i.e., the number of days since January 1 of the same year. (January 1 is Julian Day 1)
x	Time of day	N	Numeric	The time of day a specimen was collected expressed as decimal hours from midnight local time (e.g. 12.0 = mid day, 13.5 = 1:30 pm)
x	Continent ocean	N	Text	The continent or ocean from which a specimen was collected

(continued)

(continued)

Concept ID	Name	Required	Type	Description
7	Country	N	Text	The country or major political unit from which the specimen was collected. ISO 3166-1 values should be used. Full country names are currently in use. A future recommendation is to use ISO3166-1 two letter codes or the full name when searching
8	State province	N	Text	The state, province or region (i.e. next political region smaller than Country) from which the specimen was collected
9	County	N	Text	The county (or shire, or next political region smaller than State / Province) from which the specimen was collected
0	Locality	N	Text	The locality description (place name plus optionally a displacement from the place name) from which the specimen was collected. Where a displacement from a location is provided, it should be in un-projected units of measurement
1	Longitude	N	Numeric	The longitude of the location from which the specimen was collected. This value should be expressed in decimal degrees with a datum such as WGS-84
2	Latitude	N	Numeric	The latitude of the location from which the specimen was collected. This value should be expressed in decimal degrees with a datum such as WGS-84
x	Coordinate precision	N	Numeric	An estimate of how tightly the collecting locality was specified; expressed as a distance, in meters, that corresponds to a radius around the latitude-longitude coordinates. Use NULL where precision is unknown, cannot be estimated, or is not applicable
3	Bounding box	N	BOUNDINGBOX	This access point provides a mechanism for performing searches using a bounding box. A Bounding Box element is not typically present in the database, but rather is derived from the Latitude and Longitude columns by the data provider

(continued)

(continued)

Concept ID	Name	Required	Type	Description
x	Minimum elevation	N	Numeric	The minimum distance in meters above (positive) or below sea level of the collecting locality
x	Maximum elevation	N	Numeric	The maximum distance in meters above (positive) or below sea level of the collecting locality
x	Minimum depth	N	Numeric	The minimum distance in meters below the surface of the water at which the collection was made; all material collected was at least this deep. Positive below the surface, negative above (e.g. collecting above sea level in tidal areas)
x	Maximum depth	N	Numeric	The maximum distance in meters below the surface of the water at which the collection was made; all material collected was at most this deep. Positive below the surface, negative above (e.g. collecting above sea level in tidal areas)
x	Sex	N	Text	The sex of a specimen. The domain should be a controlled set of terms (codes) based on community consensus. Proposed values: M = Male; F = Female; H = Hermaphrodite; I = Indeterminate (examined but could not be determined; U = Unknown (not examined); T = Transitional (between sexes; useful for sequential hermaphrodites)
x	Preparation type	N	Text	The type of preparation (skin. slide, etc). Probably best to add this as a record element rather than access point. Should be a list of preparations for a single collection record
x	Individual count	N	Numeric	The number of individuals present in the lot or container. Not an estimate of abundance or density at the collecting locality

(continued)

(continued)

Concept ID	Name	Required	Type	Description
x	Previous catalog number	N	Text	The previous (fully qualified) catalog number of the Cataloged Item if the item earlier identified by another Catalog Number, either in the current catalog or another Institution / catalog. A fully qualified Catalog Number is preceded by Institution Code and Collection Code, with a space separating the each subelement. Referencing a previous Catalog Number does not imply that a record for the referenced item is or is not present in the corresponding catalog, or even that the referenced catalog still exists. This access point is intended to provide a way to retrieve this record by previously used identifier, which may used in the literature. In future versions of this schema this attribute should be set-valued
x	Relationship type	N	Text	A named or coded valued that identifies the kind relationship between this Collection Item and the referenced Collection Item. Named values include: "parasite of", "epiphyte on", "progeny of", etc. In future versions of this schema this attribute should be set-valued
x	Related catalog item	N	Text	The fully qualified identifier of a related Catalog Item (a reference to another specimen); Institution Code, Collection Code, and Catalog Number of the related Cataloged Item, where a space separates the three subelements
x	Notes	N	Text	Free text notes attached to the specimen record

Printed in the United States
By Bookmasters